River Basin Restoration and Management

WATER AND ENVIRONMENTAL MANAGEMENT SERIES

River Basin Restoration and Management

Edited by Avi Ostfeld and John M Tyson

IWA Publishing

Published by IWA Publishing, Alliance House, 12 Caxton Street, London SW1H 0QS, UK
Telephone: +44 (0) 20 7654 5500; Fax: +44 (0) 20 7654 5555; Email: publications@iwap.co.uk
Web: www.iwapublishing.com

First published 2005

British Library Cataloguing in Publication Data
A CIP catalogue record for this book is available from the British Library

Library of Congress Cataloging- in-Publication Data
A catalog record for this book is available from the Library of Congress

Disclaimer

ISBN: 1 84339 510 X
ISSN: 1476-1785

Contents

Preface

Over 1,800 international delegates congregated in Marrakech, Morocco for the 4th IWA World Water Congress in September 2004. This book is the result of two workshops which took place at Congress: *The Restoration of Degraded River Basins* and *River Basin Management Using machine Learning.*

The objective of the *Restoration of Degraded River Basins* workshop was to share experience on the institutional, policy, and public participation elements of restoration programmes; the 'soft' issues surrounding restoration of a degraded river basin and the development of the river basin plan.

The workshop was held in recognition of the enormous world-wide interest in how river restoration might be brought about. It is both interesting and sad to note that each of the five papers and presentations showed that human kind gives scant regard to the health of a river system until it has been damaged by human activity. In all cases an essential part of the route to improvement is the education and involvement of the public. It is evident that 'top-down' approaches do not work whereas partnerships with the local communities have a real chance of success.

The purely institutional issues are difficult and complex – perhaps no more so than on the Danube where 18 countries have territory within the Danube basin. The papers illuminate the fact that, although the requirements of basin organisations are similar, there is no model solution for universal application. Each restoration programme is more unique than similar as each must relate to: problems, objectives, resource users, political context, economics, and regulatory overlay. Key indeed is 1 broad support for action; 2 well defined objectives resulting from breadth and depth dialogue amongst stakeholders and 3 an institutional mechanism (however formal or informal) for commitment, focusing, coordination, and reporting.

The *River Basin Management Using Machine Learning* workshop highlights and compares the two different approaches to watershed management: the physically based modeling approach relying on the system physics versus the data driven modeling approach based on exploring the system "data behavior".

The workshop is motivated through the recent rapid advance in information processing systems which pushed the hydrological research community to explore the possibilities of using intelligent systems aimed at evolving models of natural phenomena automatically. This is the discipline of machine learning (ML) which is the study of computer algorithms that improve automatically through experience. Applications of machine learning range from data-mining programs that discover general rules in large data sets, to information filtering systems that automatically learn users' interests. Since the last decade machine learning techniques in both hydrology and water resources engineering are gaining an increasing interest.

This workshop assembles professionals representing a mixture of both approaches. The objective is to highlight through case studies the two attitudes and to promote through that a fruitful discussion and brainstorming.

Avi Ostfeld and John M Tyson
December 2004

The Restoration of Degraded River Basins

Restoration and Management of Degraded River Basins - The Alexander River Case Study

A. Brandeis*

*Architecture, Urban & Regional Planning and Project Management Ltd., 36 Eshkol St., Hod Ha-Sharon 45343,
Israel
(E-mail: *amos-br@inter.net.il*)

Abstract The Alexander River Restoration Project was examined as a case study for the management of a severely degraded river basin. The River Administration, established in 1995, deals with the restoration of a very complicated cross-border river which suffered from heavy pollution. The organisational concept demonstrates a coordinating, voluntary, non-institutional and non-statutory administration relying on existing organisations. Based on an interdisciplinary master plan, most pollutants were removed, seven parks were developed, the huge turtles and the ecosystem were protected, and environmentally friendly drainage methods were introduced. Everything was done with full public participation. Above all, the administrative system was successful in using the polluted river as a bridge between the Israeli and Palestinian neighbours who live along a river which knows no political borders. The unique cooperation survived even through the hard times. The conclusions indicate the importance of an entity totally devoted to the management of the basin as well as the existence of a genuine leader of the process. The comprehensive and interdisciplinary approach in combination with a practice-oriented approach, was found to be an advantage. It was found useful to have an ongoing, structured process dealing with different projects simultaneously and, in particular, to start with a demonstration project presenting uniqueness and high chances of success.

Keywords Alexander River, Comprehensive Project, Comprehensive Approach, Degraded River Basins, Environmental Projects, River Restoration, River Restoration Planning and Management.

Introduction

Degraded rivers and river basins can be found all over the world. Enhanced human activity has severely affected river basins, especially in recent decades. The basis for management and restoration of degraded rivers is the understanding that rivers and river basins differ in many basic aspects. They differ in size, location, hydrological aspects, water qualities, water quantities, ecological issues, landscape features, the nature of their surroundings, the culture of the people who live along them, their impact on the river, and many more. Even different sections of the same river are very varied. Nevertheless, all rivers have several common features: they are always the lowest point in their surroundings; they always form part of a basin which drains the water to them; they are long and narrow; they have a relatively long surface area because they are longitudinal; they are continuous, despite the fact that they have varied sections; they are always in a situation where upstream has an impact on

 River Basin Restoration and Management edited by Avi Ostfeld and John M Tyson. ISBN: 184339510X. Published by IWA Publishing, London, UK.

downstream and *vice versa*; they are very sensitive and fragile; they have an inherent ability, both rapid and effective, of recovery and rehabilitation; and many other properties.

Thousands of River Restoration Projects have been planned and carried out all over the world, in an attempt to improve the situation of degraded rivers and to manage their basins. They focus on aspects specific to the particular rivers as well as aspects which all rivers have in common. The projects also differ in capacity, approach, key issues, etc. As if this were not complex enough, we must add the intricate task of defining criteria for the success of a project, and we must determine at which stage we can say that a degraded river or river basin has been genuinely restored.

The Alexander River Restoration Project, launched in 1995, is an interesting example of a comprehensive restoration project of a severely degraded river. The project deals with many aspects which might be relevant for other, similar cases of degraded river restorations. The project won six awards, among them the 2003 International Thiess Riverprize, awarded annually in Brisbane, Australia. The main criterion for winning the prize is the ability to demonstrate progress and achievement in affecting real river management outcomes (Nelson, 2004).

The Alexander River flows through agricultural land in the northern part of the Tel Aviv metropolitan area in Israel's central district. The river actually crosses borders, since it flows from the Palestinian City of Nablus in the Samarian Hills up to the estuary of the Mediterranean Sea, situated north of the city of Netanya in Israel. Nature, scenery, several villages and extensive agricultural regions adorn its 32 km (20 miles) stretch. The river basin's size is about 550 km^2 (136,000 acres). Small quantities of fresh water flow year-round along about half its stretch. The water is less than 1.5 meter (5 feet) deep, and in some sections no more than 20 cm (8 inches). The width of the water surface is between 40 meters (45 feet) and 2 meters (2 feet). In the winter floods, which occur 1 - 5 times a year, the depth in certain sections can reach up to 8 meters (9 feet) for a few hours. In the past, the river's ecosystem was very rich, especially because of the relatively mild Mediterranean climate. The river is well known in Israel for the huge Nile soft-shelled turtles which survived in the river. These turtles are up to 120 cm (4 feet) long and weigh about 50 kg. They used to live in all Eastern Mediterranean rivers between Turkey and Egypt, but only in the Alexander River did one such community survive.

Over the last 50 years, the river basin and the river itself were severely degraded. The natural water sources of fresh water were captured and used for domestic and agricultural use. Large quantities of raw sewage and effluent were dumped into the river. About 100 sources of pollution were identified. Part of the pollutants were in Israel while others were in the Palestinian territories, between Nablus and Tul Karem. The ecology knows no political borders, even in eras of turmoil in the Middle East. The sewage from the Palestinian side continues to flow beneath the wall separating between the Israelis and Palestinians, while the mosquitoes arising from the river fly over the wall. The river became a sewage channel. Most of the natural ecosystem disappeared and the water in the river became a black and stinky source of nuisance. The degraded river attracted mosquitoes and the river banks were sprayed to avoid river bank vegetation. Flood damage was treated according to the most efficient and cheapest methods, disregarding all environmental aspects. A few meanders were shortened and concrete was used to build dams and to avoid erosion. Given its situation, the local people, most of whom had learned to swim in the river in the recent past, turned their back on the river. The river and its surroundings became the backyard of the area. Nobody wanted to get close to the waste along the sewage river. The river became an area unknown to the younger generation.

Map no. 1 - Location of the Alexander River Basin

Photo no. 1 - The Alexander River Before Restoration (1996)

Photo no. 2 - The Alexander River After Restoration (2004)

The organisational and institutional concept of dealing with the degraded Alexander River Basin

The Alexander River Restoration Administration was founded in 1995 with the aim of restoring the Alexander River and has been working tirelessly ever since. The Administration consists of about twenty public and state entities at the local, regional and national levels. The leading entities of the Administration are the Ministry for the Environment, The Jewish National Fund (JNF-KKL), and the local Regional Council and Drainage Authority. The managing concept is to establish an administration which is a voluntary, non-institutional and non-statutory body. The main principle behind structuring this unique entity was to rely on existing organisations and institutions. When necessary, the Administration's official backbone, namely the Emek Hefer Regional Council and the Sharon Drainage Authority, were consulted. The method chosen to manage the degraded basin and the restoration project was to maintain a very simple, small and almost virtual organization, with no offices or properties, to serve as the coordinator between the existing entities.

Results and discussion - the scope of work and achievements of the Alexander River Restoration Administration

The Establishment and Activity of the Alexander River Restoration Administration.

The Alexander River Restoration Administration is an innovative entity, which has been working intensively and continuously for the last nine years. It is no small feat for a non-institutional and voluntary administration to have the ability to work for years on a very complex issue, such as the relatively large-scale restoration of a cross-border river, in a very difficult geo-political situation, encompassing many different fields, aspects and river sections. The very existence of the Administration and its *modus operandi* are an important achievement in itself. The numerous key on-the-ground outcomes that can be clearly demonstrated, and the large progress in the restoration project achieved over the past few years (during a time that the main focus in the Middle East is most certainly not on ecological issues) can be considered as a success. Even more significant is the fact that the organisational capacity of the River Administration and the motivation of its leaders have succeeded - besides making major progress with the restoration of the river itself - in using this issue as a unique tool to restore hope to the Israeli and Palestinian neighbours. The hope lies in the fact that if Israelis and Palestinians can cooperate on healing an abused river, then they may be able to cooperate on other issues as well, and live together peacefully in the future.

An Overall Master Plan for the River Basin, as the Main Basis for Implementation Programs.

The first phase of the Alexander River Restoration Project involved the preparation of a comprehensive and interdisciplinary master plan for the whole river basin (Brandeis, 1996). The plan's four main components are: goals; strategy; a general land use plan; and recommendations for its implementation (an action plan). Major conflicts between disciplines were addressed in working groups, workshops and meetings. The master plan attempted to attain the delicate balance between ecology, flood defense, economy, landscape planning, tourism, etc. The plan includes a detailed action plan, with 135 projects and activities required to implement the plan, in a wide range of areas. Priorities were set for each one of them, and an estimation of the cost of each project was added. This plan was, and still is, the basis for the on-the-ground restoration works.

Outline Scheme as an Important Legal Basis for Implementation and Economic Feasibility.

Three outline schemes for the various parts of the river were prepared in 1997 – 2000 (Brandeis, 2000). These plans were designed to legally establish the goals and principles of the master plan in statutory outline schemes, and to protect the open spaces along the river ("The River Strip") against the heavy pressure exerted by developers. They are currently in advanced stages of approval. The plans enable entrepreneurial development of recreational and leisure activities. They will also be a source for financing maintenance. The outline schemes called for innovative statutory tools to create the delicate balance between restoration and development in an area where the open spaces are under real threat by developers. These tools, which were approved by the National Planning Committee of Israel, subsequently received an award. Similar plans dealing differently with the land use issue were prepared in Israel for the most polluted river in Israel – The Kishon River (Brandeis, 2001), and for the Yarkon River which flows through the city of Tel Aviv (Rahamimoff, 1998).

Multidisciplinary and Comprehensive Implementation of the River Restoration.

Despite the River Restoration Administration's non-institutional character, it has run a large variety of projects since 1998, which are of a truly diverse nature and cover different sections of the river.

The "Demonstration Project" is the *pièce de résistance* of the river restoration. The project involves the scenic, visual, ecological and drainage restoration of a river section of approximately 750m (half a mile). This part of the river was chosen due to its run-down condition prior to the project's development and due to its prime accessibility and exposure to the adjacent highway. The essence of the project was to develop an attractive river park intended to serve, on the one hand, as an illustration of the potential inherent in the restoration of river areas and, on the other, as an experimental project from the point of view of many aspects. The planning and implementation are based on international experience acquired from running projects of this type, like the Skerne River Project in Denmark (Hansen, 1996) and the Cole in England (Vivash, 1999) The project attains a pre-agreed balance between drainage, ecological, scenic, tourist, maintenance, implementation and other considerations. The project was completed in 2002 (planned by landscape architect Amit Segal).

A few more river parks have been developed over the last years, along different sections of the river and its main tributary. The nature of the development differs according to the setting and natural resources of each park. One of the parks was developed during 2002, upstream, in an area which lies between four villages, all with a relatively low socio-economic status. The project also included paths running from the villages to the river. This river section, which is about 3 km (2 miles) long, was chosen in order to strengthen the relationship between the river and the local community, which was using this section as if it were its own backyard, and treating it accordingly. Community work, especially with the youth, was introduced in addition to the physical development works.

The removal of pollutants is the main issue handled by the Administration. Three main sources have been eliminated since the Administration was established. The Ministry of the Environment, a senior partner of the Administration, played a key role in this important achievement. The public pressure, caused by the publicity given to the Project, also helped.

More projects have been carried out to protect the rare, huge Nile soft-shell turtles that live in the river and to rehabilitate the natural ecosystem (for example, the building of riffles and fish-ladders, based on experience from Denmark and Scotland). One component of the project was called 'Saving Alex', referring to the turtles which are unique to this river. Basking areas and protected egg-laying areas were developed for the turtles.

Maintenance, River Operation and Monitoring.

Maintenance is one of the most difficult issues in river management. All the Administration's partners like to be involved in the interesting and prestigious work of developing new projects. Not many of them like the hard work of daily maintenance. One of the reasons that the Alexander River has been successful is due to the fact that the chairman or manager of almost all the organisations responsible for maintenance is the same person. This is why the collaboration in work and budgeting between the Regional Council, the Drainage Authority, the local Water Council and others is very good. The Administration has a sub-committee which deals with the issues of maintenance, adopting environmentally-oriented methods used in other countries. Most of the maintenance work is done by the Drainage Authority and private contractors hired by the Administration.

The Role of the Project in Establishing Real Cooperation between Israelis and Palestinians and Solving Ecological Problems that Know No Geo-Political Borders.

The partnership between the Israelis and the Palestinians and the cooperation between them on the river restoration is a unique achievement of the Administration. There are endless tales about the bridge built between the people living along the polluted river, who cooperate as neighbours sharing common ecological problems, despite the troubled region and the era of turmoil they live in. The excellent cooperation achieved on a personal level, in the face of the ecological and health problems of the polluted river, offers hope that the River Restoration Project can serve as a bridge between the Israeli and Palestinian neighbours. And indeed, this hope seems to have materialised, in part at least, given that the Israelis and Palestinians have continued to cooperate on this project during the last years, even though a four-meter high security wall has been built between the Israelis and Palestinians, and the tension in the region has been growing higher.

The sewage flowing in the Nablus Stream is the main source of pollution in the Alexander River. The sewage originates in the city of Nablus, located within the Palestinian Authority, and flows adjacent to the town of Tul Karem. The Nablus Stream crosses the border into the State of Israel, where it joins the Alexander River. The Nablus Stream has approximately 70 sources of pollution, such as sewage arising from urban areas, villages, refugee camps and industries. During the olive harvest, 26 olive mills pollute the river by dumping olive waste (the residue remaining after olive-oil production) into the channel. The river is further polluted by the stone grindings from the many marble-processing factories situated along it. During the olive harvest, the river suffers from an extremely high organic substance concentration (equivalent to the sewage of a city of 300,000 inhabitants), which leads to severe fish mortality and colours the water completely black. The stone grindings precipitate along the entire river, which give the water a grey-whitish color all year long.

In 1996, while planning the Alexander River, the head of the Emek Hefer Regional Council, who also serves as the chairman of the River Administration Board, initiated cooperation with the neighbouring Palestinians. The environmental and health problem provided the impetus to promote neighbourly relations, based on local cooperation rather than dialogues between states. Two agreements, of a unique nature, were signed by both sides with the aim of cooperating and solving the problem.

A joint Israeli and Palestinian planning team prepared a survey and a plan to build a joint purification plant along the border. This plant will hopefully be built as soon as the situation in the region improves. It will treat all the sewage from the Palestinian town of Tul Karem, and all the sewage diverted from the river bed on the Palestinian side. The plant will be operated jointly, and the treated water will be used for irrigation on both sides and for the restoration of the Alexander River. The plan was jointly approved in 1999.

The German government undertook to finance the project at an overall cost of approximately USD 25 million. A German engineering company prepared a feasibility plan and, on the whole, cooperation is proceeding positively. In 2004, construction works began in Tul Karem. The aim is to restore the cities' sewage ponds and to connect them to the regional sewage system, instead of discharging the sewage to the river. The works are being performed by a Palestinian contractor with German funding.

Since the Alexander River was undergoing intensive restoration, a decision was made in the year 2000 that the River Restoration Administration would implement an emergency solution, rather than waiting 4 - 5 years to set up the purification plant (provided that there are no difficulties or delays of any kind).

This solution will subsequently be integrated into the permanent solution. The Israeli team presented the plans for the emergency project to their Palestinian neighbours.

During the initial stage, the sewage is pumped from the riverbed, treated and returned - with a much improved quality - to the riverbed. During the second stage, most of the water will be utilised for irrigation in both Palestinian and Israeli fields (Brandeis, 2003). The project was constructed between 2000 - 2002, based on an investment of about USD 4.5 million by the Administration. In September 2002, Palestinian representatives were able to attend the opening ceremony of the project.

Photo no. 3 (left) - The polluted riverbed on the border between the Israelis and the Palestinians. Photo no. 4 (right) - "The Emergency Project" for the treatment of the Nablus Stream Sewage

The Project's Role in the Change of Attitude to Rivers in Israel and Within the River Basin.

Thanks to the extensive publicity it has received, the Alexander River Restoration Project, as the leading river restoration project in Israel, has succeeded in making a significant contribution to changing the negative attitude to rivers in Israel which had prevailed for the last fifty years. The Administration has also developed its own professional status so that most of its members seek professional advice.

Real Public Participation.

Since the inception of the project, the principles of transparent and accountable planning were among its top priorities. The importance of public participation as a major factor for success has been studied in many projects (Riley, 1998). The plans were designed in an intensive process of collaboration with the public. 21 out of 22 villages that possess land within the outline scheme, approved it. Public hearings were held at different planning stages. The scheme was submitted to the planning authorities, in order to offer the public the opportunity to comment on and/or oppose it. Children and students have taken part in various activities such as planting trees along the river, documenting the restoration tasks, returning fish to the river, taking part in guided tours along the river, preparing assignments, etc. Each school in the area has a special curriculum for its students pertaining to the Alexander River Restoration process. Students have been given special lectures and tours, and they prepared a wide variety of impressive papers, songs, etc. In 2004, a special school program was prepared for 4th grade students (Raveh, 2004). This was only possible thanks to the International Riverprize awarded to the project in 2003 (100,000 Australian Dollars). The restoration of the river has become a very important component of the vision of the region as a whole.

Openness, Accountability, and Regular Reporting of Achievements.

The project has acquired a high degree of openness and accountability, and enjoys regular media coverage. Since the project won the International Riverprize, more than 60 articles have been published and many interviews have been held. Among them the articles published in the New York Times (Rosenberg, 2004) and in "Water & Wastewater International" (Rosenberg, 2004). Since the inception of the project, public meetings have been held frequently. A relatively small-scale 'River Festival' with 7,000 participants was launched in 2002 (based on the idea learned from Brisbane, Australia), and in the same year, a musical parade took place along the river for the first time. A film was produced and distributed to all the schools and villages in the region. A web site was launched in 2002 (www.alexander-river.dynu.com).

Budget and Effectiveness of Use at Available Resources.

The overall budget for the Alexander River Restoration Project has exceeded USD 12 million in the space of a few years. This is much higher than any other river budget in Israel. The Administration works in a precise and efficient way, according to a very strict timetable. It has a five-year action plan, which is updated every year according to the progress of the work and budget. The Administration's organisational structure allows for minimal current expenses. The River Administration's only paid employee is the project manager (part-time). The Administration makes use of existing administrations, such as the Drainage Authority and the Regional Council. The low current expenses mean that there is more money for allocation to the actual restoration activities, which is highly appreciated by the members.

Unique Cooperation.

The Administration is an example of unique cooperation in all aspects and at all stages of decision-making, planning, and implementation. It is an example of cooperation between about twenty public and state entities at the local, regional and national levels. All entities, which are normally involved in disputes with each other, actually agreed to collaborate and help. All disputes and personal interests were put aside. All the entities approved the master plan and the annual programs. The plan was also approved by the planning committees and the government of Israel.

Conclusions

The aim of restoring and managing a degraded river basin is a very complex and challenging mission. Even though rivers and river basins differ very much in their characteristics and the degree of degradation, general conclusions may be drawn from case studies. Conclusions from the Alexander River Restoration Project can be studied and implemented in other projects - however, the adoption of the conclusions must be done in a critical way, because there is no general prescription for the restoration of a river, nor will there ever be. Each river is very different with its own unique planning climate.

An Organisational and Institutional Entity Devoted to the Restoration Management of a Degraded River Basin.

It seems that a basic condition for success is the establishment of an entity focusing solely on the issue of the restoration and management of a degraded river basin. This entity can be a voluntary and non-institutional administration like the Alexander River Restoration Administration, an institutional river authority, or any other organisational structure. The advantage of an entity like the Alexander River Restoration Administration is the fact that it relies on existing organisations and is therefore very efficient. The disadvantage is that the entity is very fragile and any shortage of funds, even temporary, may bring the project to an end. An institutional authority is more stable and usually has also more capacity and sometimes even a legal status of its own with power of enforcement. Such an authority has relatively high overheads, and might be in conflict with other entities, because its power is sometimes in conflict with that of other organisations.

A Comprehensive Interdisciplinary Approach.

To restore any river, it is useful to study all the relevant aspects in all fields and to prepare a combined strategy (Brookes and Shields, 1996). Experts from all disciplines should be involved in the planning and implementation of projects, however decisions must be based on a synthesis and a comprehensive approach which forms a delicate balance. After four years of planning and managing river restoration projects (figure no. 2), the problem seemed to be two- dimensional and relatively simple (Brandeis, 2001). After nine years, it appears to be three-dimensional and very complicated (figure no. 3). It is like solving a Rubik's Cube: it entails many aspects; it is three-dimensional and complicated; if you make one wrong move you have to retrace your steps; and you know that other people have solved the problem before you.

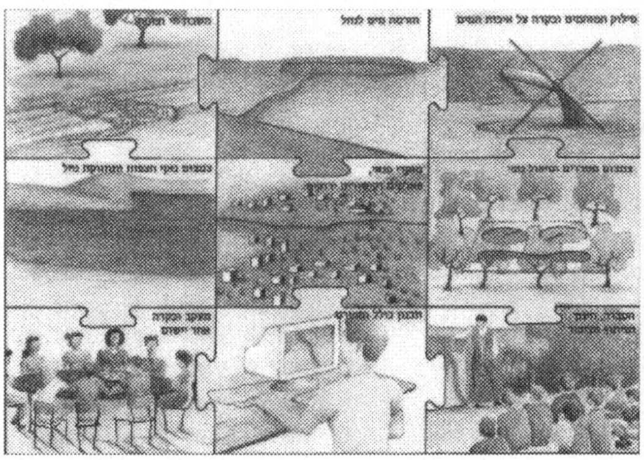

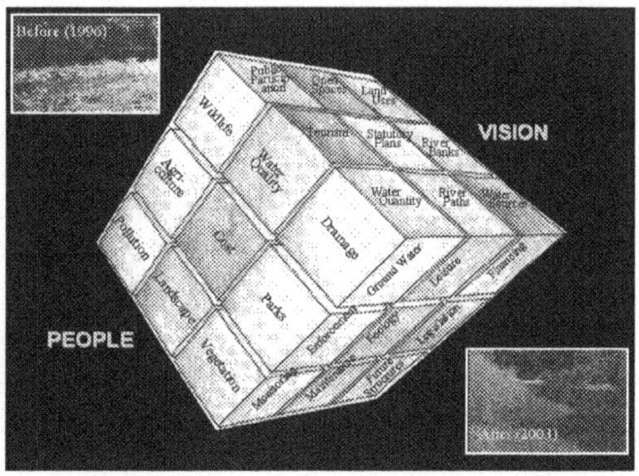

Figure no. 2 (left) - Conclusions after four years: The problem of restoring a degraded river basin is like solving a complicated puzzle. Figure no. 3 (right) - Conclusions after nine years: The problem of restoring a degraded river basin is like solving the three dimensional Rubik's Cube.

The Right People and the Vision. To solve a River Restoration Project you need two essential components, like the two hands you need for solving the Rubik's Cube: the people and the vision. Only if you have a leader for the process with power, standing, motivation, skills and sufficient time, will you succeed in achieving effective collaboration of all the relevant partners, which, in turn, will lead to a successful project. A clear vision based on the realistic definition of long- and short-term goals is the second condition for a successful project.

Practice-Oriented Approach. The partners of a restoration project, like the partners of the Alexander River Restoration Project, will usually appreciate the promotion of projects implemented on the ground, rather than the preparation of surveys, researches and reports.

Ongoing Conditions Process. It is very important to have a clear, long-term working plan, which may be examined and altered, but not interrupted. Success relies very much on a continuous momentum.

Structured Planning and Implementation Process. The process adopted by the Alexander River Restoration Project can serve as an example of a structured process - a comprehensive master plan, outline scheme, implementation of projects and monitoring. *Ad-hoc* work is usually not recommended.

Different Projects Simultaneously. From the experience of the Alexander River Restoration Project, all efforts should not be concentrated on a single project or on a single river section. Work on different kinds of projects, in different river sections, meets the interests of the project's various partners and helps to continue the momentum of the project.

Small Successes are Better than Large Failures. It is better to start with small projects with guarantied success rather than fail with an overly ambitious project.

A Demonstration Project. One clear conclusion of the Alexander River Restoration Project and of projects in England (Vivash, 1999) and in Denmark (Hansen, 1996), is that it is advisable to commence with a project to demonstrate the potential and also to serve as a pilot study.

Public Participation. Genuine public participation is instrumental in achieving a better project and in enhancing the relationship between the people and the river basin.

Uniqueness. Unique and interesting features or projects can help to gain more support and funding.

A Broad Approach – Meeting Other Goals Too. It is possible to meet other goals, too, extending beyond the river restoration itself: e.g. regional goals, collaboration between people/enemies, etc.

A Bridge between People and Enemies on a Cross Border River. A river restoration project can be the bridge to overcome major conflicts, because ecology knows no political borders and the deterioration of the ecosystems causes terrible damages to all the people in the region. The Alexander River Restoration Project succeeded in raising some hope for a better future for the people who live along the river.

Budget, Patience and Luck. If you don't have these, don't even attempt to restore a river.

References

Brandeis A. (1996), *Alexander River Restoration Master Plan* (not published).
Brandeis A. (2001), *Kishon River Masterplan* (website: www.kishon.org.il).
Brandeis A. (2003), *Alexander River Restoration Administration 2002 Report* (not published).
Brookes A., Shields D. JR. (ed.) (1996), *River Channel Restoration*, Wiley J.& Sons Ltd, Chichester, England.
Nelson, (2004), *The 2004 Riversymposium* Brisbane (Brochure and Website: http://www.riverfestival.com.au/).
Hansen O.H. (ed.) (1996), *River Restoration - Danish Experience and Examples*, National Environmental Research Institute, Denmark.
Rahamimoff A.. (1995), *Yarkon River Master Plan* (website: http://www.yarqon.org.il).
Raveh R. (2004), The Alexander River – School Program for 4th Grade, The Alexander River Administration.
Riley A. L. (1998), *Restoring Streams in Cities*, Island Press, Washington DC, USA.
Rosenberg, (2004), *Water &Wastewater International, May 2004 Issue.*
Vivash R. (main author), Murphy D. (assistant), Janes M., Holmes N., Haycock N. (ed.) (1999), *Manual of River Restoration Techniques*, The River Restoration Center (RRC), Sisloe Campus, Beds, England.

The Development of a Catchment Management Agency for the Breede River, Western Cape, South Africa

G E McConkey*, W D Enright*, J A Roberts*, MAR Khan*,

*Department of Water Affairs and Forestry, Private Bag X16, Sanlamhof 7532 , Cape Town, South Africa.
gem@dwaf.gov.za

Abstract The development of a Catchment Management Agency for the Breede-Overberg Water Management Area is the result of the deliberations of the Breede-Overberg Catchment Management Agency Reference Group, assisted by the Department of Water Affairs and Forestry: Western Cape, South Africa. Individuals, organisations and groups who participated in this process gave much time and energy to sharing information, learning more about integrated water resource management and the challenges facing others, and finding consensus positions on difficult issues. As in all other Water Management Areas in South Africa, progress towards the establishment of new water management institutions, in general, and a Catchment Management Agency, in particular, involves a radical change in the way in which water users and interest groups relate to water resources. Not only is the role of the Department of Water Affairs and Forestry changing but water users and interest groups are grappling with new opportunities and demands on their time, energy and knowledge. Members of the Breede-Overberg Catchment Management Agency Reference Group have engaged with these issues and are committed to continued growth and engagement after the establishment of a Catchment Management Agency for the Breede-Overberg Water Management Area.

This paper describes very concisely the complex process of establishing a Catchment Management Agency as it is done in the South African situation in terms of the National Water Act 1998 (Act 36 of 1998).

Keywords institutional development, river basin management, water resource management

Background

The promulgation of the National Water Act, 1998 (Act No. 36 of 1998) set in motion a process of fundamental change in the way in which water resources are managed in South Africa. One of the most significant aspects of this process is the establishment of new water management institutions that will allow water user and interest groups to participate in the management of their water resources. The most significant of these institutions is the Catchment Management Agency (CMA).

The National Water Act makes provision for the creation of CMAs to manage the water resources within specific Water Management Areas. In his capacity as custodian of the nation's water resources, the Minister of Water Affairs and Forestry, through the Department of Water Affairs and Forestry (DWAF), has divided South Africa into 19 Water Management Areas (WMAs).

The Breede-Overberg WMA is one of these, and water user and interest groups in this WMA took up the challenge to develop a "Proposal for the establishment of a CMA for the Breede-Overberg WMA"

(hereafter referred to as "the Proposal").

The Proposal was prepared in terms of Section 77 of the National Water Act 36 of 1998, which requires that proposals include the following:

1. a proposed name and a description of the proposed water management area of the agency;
2. a description of the significant water resources in the proposed water management area, and information about the existing protection, use, development, conservation, management and control of those resources;
3. the proposed functions of the catchment management agency, including functions to be assigned and delegated to it;
4. how the proposed catchment management agency will be funded;
5. the feasibility of the proposed catchment management agency in respect of technical, financial and administrative matters; and
6. an indication whether there has been consultation in developing the proposal and the results of the consultation.

The contents of the Proposal was the result of deliberations by the Breede-Overberg CMA Reference Group, assisted by a support team that functioned under the guidance of DWAF: Western Cape.

In the Proposal, a summary of the participation process was given. This shows how water user and interest groups became involved in the proposal development process, how the Breede-Overberg CMA Reference Group was established, how it functioned, the capacity building and empowerment activities that were undertaken in the WMA, and representation of groups and perspectives. This process concludes with comment from stakeholders regarding the participation process.

The proposal then focuses on the assessment of the feasibility of the proposed CMA. This section includes a description of the Breede-Overberg WMA, a description of the proposed delegation of functions, a description of the proposed institutional evolution and an assessment of organisational viability, an assessment of financial viability, an assessment of social viability and a summary of the requirements for a viable CMA.

The proposal then concludes with an implementation plan for the establishment of the proposed CMA. This includes a description of the activities, responsibilities, timeframes and resources for the first year of the life of the CMA.

Requirements for a viable Breede River CMA

Institutional and technical viability

In terms of the institutional and organisational evolution of the CMA, one should note that this proposal acknowledges the need for flexibility and caution during the development of the CMA. This is due to uncertainties in the policy environment (e.g. aspects of the Pricing Policy, such as a Waste Discharge Charging System, still needs to be developed fully), the complexities of restructuring of DWAF and establishment of Water User Associations (WUA), and the need to ensure that due diligence is followed when deciding on the readiness of the CMA and WUA's to take on water resource management functions. This Proposal allowed sufficient flexibility and allowed for the participation of all key role players to ensure responsible organisational development.

The work undertaken to date by DWAF: Western Cape to establish functional organisation of water resource management activities in terms of WMA boundaries, including on-going work on tracking the costs of such activities, will stand the new CMA in good stead. Transfer of functions, secondment and transfer of personnel, provision of operational assistance to the new CMA and management of financial flows between the CMA and DWAF will all be aided by this work.

Schedule 4 of the National Water Act 36 of 1998 provides broad guidance regarding the functioning of the future CMA Governing Board as well as sound financial management and reporting arrangements. The CMA Governing Board will have to entrench these guidelines in a constitution that reflects the goals of the National Water Act 36 of 1998. It will also have to develop practical operational policies and procedures, including a human resources policy and financial and information management systems. Outside expertise may have to be contracted in to assist the new CMA to develop such policies and systems. Adoption of existing DWAF systems will not be an option, as they have been designed with different purposes in mind. However, CMA operational systems, particularly information and financial management systems, have to produce the outputs required by DWAF for its monitoring and auditing purposes. Existing DWAF operational practice and the requirements of the National Water Act 36 of 1998 will, therefore, need to be considered when such new systems are designed.

Finally, there is sufficient expertise in and near the Breede-Overberg WMA to assist the new CMA during its development. There are strong organisations and institutions on which the CMA can draw for technical support while undertaking operational tasks for which it has been given responsibility.

Financial viability

The process of the registration of water uses and water users should be complete by the time the Breede-Overberg CMA is established. The National Water Resources Strategy should also be in place, giving an official estimation of allocable water in the Breede-Overberg WMA. This will give the new CMA clear indication of its revenue base.

The new CMA will have to remain sensitive to the needs of water users to ensure that water resource management charges are paid. The proposed representational/governance structure should allow all water users to have a stake in the work of the CMA and give them insight into the need for activities for which they are being charged.

A system of differential charges should be established as soon as possible. This should include differential charges for irrigation use based on assurance of supply and differentiation between charges for groundwater and surface water. The former is particularly important, as there is concern about the ability of some users in the agricultural sector (which is the largest user sector and, consequently, the largest potential source of revenue for the CMA) to pay. Phasing in of the water resource management charge for emerging farmers needs to be based on an agreed definition of "emerging farmer" that is practically implementable.

Also to be included in the system of differential charges should be differential charges for out-of-WMA transfers for irrigation and municipal purposes.

The annual budget for year 10 of the life of the CMA is, in general, affordable and compares favourably with annual turnover figures of a large local WUA such as the Central Breede WUA. However, care will have to be taken to ensure affordability to individual water users. Differential charge systems and careful consideration of the joint impact of water resource management charges and water resource development charges on water users will be important in this regard.

Social viability

Public participation and capacity building support should be maintained after submission of the final *Proposal for the establishment of a CMA for the Breede-Overberg WMA* and after establishment of the CMA. Once it has been established, the CMA should take over responsibility for public participation and capacity building support. Tying such support to planning initiatives such as catchment management planning initiatives will ground them more and increase the enthusiasm of participants.

Establishment of effective cooperative governance relationships will be of crucial importance to the work of the future Breede-Overberg CMA. Strong relationships with District and Local Municipalities within the Breede-Overberg WMA are of particular importance given the growing developmental mandate of local government in South Africa. Provincial and national government departments are investigating ways of facilitating implementation of their mandates at local government level within well-conceived development frameworks. Integrated development planning at local government level is likely to grow in importance, not only because, form a CMA's perspective, it produces Water Services Development Plans in which domestic (includes industrial) water needs are specified. It is through these and provincial planning processes that the social and economic development needs for which strategic resources, such as water, need to be managed.

The CMA should not attempt to take responsibility for socio-economic planning; it should work closely with those organs of state that are responsible. The CMA will have a clear mandate to manage the water resource, as defined by the National Water Act 36 of 1998, and will have to engage actively with a range of spheres and departments of government. By drawing all water user and interest groups into its work, it can contribute to the on-going process of aligning resources, needs and plans in the interests of all of the people of the Western Cape, particularly those in the Breede-Overberg WMA.

Implementation plan for CMA establishment

After submission of the final "Proposal for the establishment of a CMA for the Breede-Overberg WMA" to the DWAF Directorate: Catchment Management for initiation of the approval procedure, a period of waiting begins. According to information received from that Directorate, a minimum period of 9 months should be allowed for the approval process. This period and the various steps in the approval process are included in the implementation programme below.

During this period, the Breede-Overberg CMA Reference Group will interact with the Advisory Committee: Governing Board (not yet established) regarding membership of the Governing Board of the future Breede-Overberg CMA. A separate proposal will be prepared in this regard, allowing members of the CMA Reference Group to clarify their thoughts about membership of the future CMA Governing Board.

Also during this period, the process of building the capacity of water resource committees and forums in the Breede-Overberg WMA can continue, linked, if possible, to planning processes such as a catchment management planning process or the process of harnessing stakeholder comment on the draft National Water Resources Strategy. DWAF: Western Cape will have to continue to take responsibility for such activities in the Breede-Overberg WMA until the CMA has been established.

Once the Advisory Committee: Governing Board has made its recommendations to the Minister of Water Affairs and Forestry regarding membership of the Governing Board of the future Breede-Overberg CMA, and once the final *Proposal for the establishment of a CMA for the Breede-Overberg WMA* has been approved, the Minister of Water Affairs and Forestry will formally establish the new CMA.

The first and most important step in the establishment process is the appointment of the Governing Board. One of the first tasks of this new CMA Governing Board would be the election of a chairperson to assume the responsibilities described in schedule 4 of the National Water Act 36 of 1998. The newly elected chairperson would have to convene a meeting between the newly appointed CMA Governing Board and senior officials of DWAF: Western Cape to establish a firm working relationship between the two parties. It may be advisable to establish an executive management committee consisting of some or all members of the CMA Governing Board of the CMA and senior DWAF: Western Cape staff officials (in the beginning it would make sense to involve all members of the CMA Governing Board, even if only to build knowledge of DWAF operations and practical water resource management challenges). The DWAF contingent should include at least the Catchment Manager: Breede Water Management Area. The CMA Governing Board and the executive management committee can draw in additional expertise (individual or from other organisations) when needed.

The first tasks of the CMA Governing Board (after election of a chairperson) would include:

- Development of a constitution based on schedule 4 of the National Water Act 36 of 1998 and serving to promote the goals of that Act.
- Development of a human resources policy (initially very broad, but with job descriptions for the first expected appointments or secondments). The services of human resources specialist might be required to assist the Governing Board in approving a policy that meets the requirements of South African labour legislation and best practice human resource management.
- Establishment of the Catchment Management Committees (CMC's). This might involve formal recognition of the existing Breede River Basin Study and Overberg Stakeholder Committee as Breede and Overberg CMCs, but the CMA Governing Board will have to satisfy itself that membership is representative at the time of CMC establishment.
- Activation of communications channels between the CMA, DWAF, CMCs, Catchment/Geographical Steering Committees, Forums, WUAs and organs of state. Input by DWAF: Western Cape will be valuable in this regard, as it has established such communications channels already. These existing channels will have to be reconfigured to serve the needs of a new and evolving statutory body, the CMA, while still serving DWAF: Western Cape's needs.
- Development of the first Breede-Overberg CMA business plan. This plan has to be completed within six months of establishment of the CMA Governing Board. The business plan may have to be completed before the first appointments are made, as a period of 6 months would be the time required to prepare adequately for the first appointments and secondments.
- Appointment of a Chief Executive Officer (CEO) or a Manager: Technical Services. Sufficient time should be allowed for preparation of a minimum network of systems and procedures before any appointments are made. The CMA Governing Board will have to decide whether to appoint a CEO immediately, or whether to make a lower level appointment to help prepare the ground for the eventual appointment (one to one-and-a-half years after establishment of the CMA) of a CEO. Although one might see the immediate appointment of CEO as allowing such an appointment to put his/her stamp on the organisation from its inception, there could be good reasons to appoint a Manager: Technical Services first. These include:

  - More time is allowed for recruitment of a CEO with wide experience.
  - The person appointed as Manager: Technical Services could be given the opportunity to learn from current DWAF: Western Cape staff and to prove his/her mettle in a range of management activities.
  - Such an appointment would open up the possibility of secondment of DWAF: Western Cape staff currently assigned to the Breede-Overberg WMA to the CMA, with the current Catchment Manager acting as CEO. Such an arrangement would allow for very close operational cooperation between DWAF: Western Cape and the new CMA, but, as indicated previously, reporting channels between seconded staff, on the one hand, and DWAF and the CMA Governing Board, on the other, would need to be thought through carefully.

- Appointment of secretarial staff: Such staff will be particularly useful in keeping records of meetings of the CMA Governing Board and an executive management committee, in typing correspondence, serving as a voice when water user and interest groups try to contact the new CMA and to undertake logistical tasks such as organisation of meetings.

  The first tasks of an executive management committee would include:
  - Development of a *modus operandi* for DWAF support that can be submitted for approval by DWAF and the CMA Governing Board.
  - Negotiation of personnel transfer and secondment options (on-going). Again, recommendations would have to be approved by both DWAF: Western Cape and the CMA Governing Board.
  - Coordination of operational support to the new CMA.
  - Development of recommendations regarding job descriptions of the first appointments and secondments to the CMA.
  - Development of recommendations regarding the first CMA business plan.

The first appointment to the CMA, whether a CEO or a Manager: Technical Services, should work closely with the DWAF: Western Cape Region, but should be based in the Breede-Overberg WMA, in a place chosen by the Breede-Overberg CMA Governing Board. Close operational cooperation with DWAF: Western Cape staff will allow this person to familiarise him-/herself with the latter's activities and to identify operational challenges.

If the first appointment to the CMA is a Manager: Technical Services, rather than a CEO, it is likely that a CEO appointment will be finalised at most 1 year after the appointment of the Manager: Technical Services. This would mean appointment of the CEO one-and-a-half years (at most) after establishment of the CMA Governing Board.

Water resource management charges will have to be collected by DWAF: WC for the first year, after which arrangements can be made for the CMA to take over responsibility (possibly through the use of agents, which could be WUAs and/or Overberg Water). After one year the appointed and seconded staff of the CMA should be able to coordinate revenue collection arrangements. Once the CMA has assumed responsibility for revenue collection, it will have to transfer funds to DWAF to cover the latter's water resource management activities in the Breede-Overberg WMA.

Internal Strategic Perspective (ISP)

The Internal Strategic Perspective is another process that is done in parallel to the formal procedures required to establish a CMA. The Internal Strategic Perspective (ISP) is a DWAF view on how it intends managing the water resources within the Breede WMA during the period leading up to the establishment of a CMA and the development of a Catchment Management Strategy (CMS) by the CMA. The ISP will provide a consistent basis for the Western Cape Regional Office (RO) of the DWAF to process requests for new water use licences.

After internal approval, the Department will obtain comment on the ISP from local authorities, Water User Associations (WUAs), other water related organizations and the public. This will take place through the established liaison structures (forums) and through normal contact with water users in the execution of its duties. The ISP document will be updated periodically until the CMA for the Breede WMA is fully functional and has prepared a CMS. All updates to the document will be authorised by the RO, and the task of managing version control will be the function of the Catchment Manager for the Breede WMA.

The ISP has been compiled by referring to policy documentation, legislation, regional planning, departmental guidelines and relevant water related studies, and is also based on interviews and communications with DWAF's regional managers, as well as staff in relevant Head Office directorates.

Conclusions

This paper indicates very concisely the complex process of establishing a Catchment Management Agency that is viable and able to manage the water environment according to an agreed Catchment Management Strategy.

The legislation insists on a very participative approach and a gradual shift of responsibilities from the DWAF to the CMA in a manner that will not hinder the services delivery of water and the protection of water resources.

There are nineteen Water Management Areas in South Africa and all of these will eventually be transformed into Catchment Management Agencies. The process of establishment is the same but there are tremendous differences in capacity, availability of water resources, economic viability and social development. The development of all nineteen CMA's will be very challenging.

References

Department of Water affairs and Forestry (2002) - Proposal for the establishment of a Catchment Management Agency- Breede River Water Management Area- Developed in terms of Section 77 of the National Water Act, 1998 (Act No. 36 of 1998) for submission to the Minister of Water Affairs and Forestry - Internal Document, Pretoria, South Africa

Department of Water Affairs and Forestry (2003) - Breede Water Management Area- Internal Strategic Perspective- Internal Draft Document, Pretoria, South Africa.

Department of Water Affairs and Forestry (1998) -The National Water Act (Act No.36 of 1998)- Govt. Gazette No 19182, Government Printer, Pretoria, South Africa

Mersey Basin Campaign: a partnership approach to river basin management

A. Wright[1] and B. Bendell[2]

[1]Research and Information Manager, Mersey Basin Campaign, City Tower, Piccadilly Plaza, Manchester, M1 4BT. (E-mail: *a.wright@merseybasin.org.uk*.)
[2] Head of Corporate Responsibility, United Utilities plc, Dawson House, Great Sankey, Warrington, WA5 3LW. (E-mail: *bruce.bendell@uuplc.co.uk*.)

Abstract The Mersey Basin Campaign was established as one of a series of initiatives, launched in the wake of riots in Liverpool, that were designed to kick start the regeneration of England's Northwest. At the time, the then Secretary of State for the Environment, Michael Heseltine called the River Mersey "an affront to the standards a civilised society should demand of its environment." It was recognised that a combination of public, private and voluntary sector action was necessary, leading to the Mersey Basin Campaign partnership being formed in 1985. This partnership has developed since its inception, with changes in the governance structure of the Campaign creating wider participation from business, academia, community and voluntary groups. The partnership of the Mersey Basin Campaign brings together disparate groups and sectors to work together towards a shared mission. Today, the Mersey and its tributaries are cleaner than at any time since the end of the industrial revolution. The transformation of the River Mersey has been supported by a combination of major investment and genuine partnership working at regional and local level.

Keywords Catchment management; partnership; private sector; water quality

INTRODUCTION

Background to the creation of Mersey Basin Campaign

In the 19th century, the north west of England became the World's first industrialised region. The rapid industrial growth created a demand for labour and thus brought about very rapid expansion of urban areas. Domestic sewerage systems were based on untreated disposal directly into rivers and sea. Manufacturing industry became established along the region's rivers and new canal system, which became the major conduits for removing and transporting industrial wastes.

By 1869 the Mersey and the Irwell, which in 1721 supported fish as a commercial industry, had become so grossly polluted that a Royal Commission on Rivers Pollution was appointed to study and report on the problem (HMSO 1874). In so far as the problems were recognised, little priority was attached to addressing them by the municipal authorities. Certainly as late as the 1980s, the Mersey was the most polluted estuary and river system in the UK. Throughout the 20th century progressive changes to legislation and institutions, including the formation of water authorities in 1974, brought about significant improvement but, even so, towards the end of the century, the region's waterways were among the most polluted in the world (MBC 1985; Kidd & Shaw 2000), and industrial decline was manifest in dereliction, poor housing and growing social problems.

These problems came to a head in the 1981 when disturbances in Toxteth, an inner-city area of Liverpool, turned into full-blown riots. In the wake of the riots, government policy began to focus on the problems of inner cities, and of post-industrial dereliction. A number of targeted regeneration programmes and projects followed, including the Mersey Basin Campaign.

The Campaign was established in 1985, with government backing and a 25-year initial life span. Its role was to address the problems of water quality and associated landward dereliction of the River Mersey and its tributaries. (It has since expanded to include the River Ribble in Lancashire). The development of the Campaign broke new ground in British administrative practice in a uniquely collaborative programme founded on a personal initiative of a government minister, Michael Heseltine, then Secretary of State for the Environment (Wood, Handley & Kidd 1999).

The objectives of Mersey Basin Campaign

The objective for this new approach was set out clearly:

'To rebuild the urban areas of the North West we need to clean and clear the ravages of the past, to recreate the opportunities that attracted earlier generations to come and live there and invest there…A Mersey Basin restored to a quality of environmental standards fit for the end of this century will be of incalculable significance in the creation of new employment.' (Michael Heseltine, Department of the Environment, 1982 pp.1-2).

He recognised, quite clearly, the interdependence of economic prosperity and quality of environment, with the Campaign conceived as a 'sustainable development' approach long before these words were in common currency. This translated into three key objectives for the Campaign, identified at the start of the initiative:

- to improve river quality across the Mersey Basin to at least a 'fair' standard by 2010 so that all rivers and streams are clean enough to support fish;
- to stimulate attractive waterside developments for business, recreation, housing, tourism and heritage; and,
- to encourage people living and working in the Mersey Basin to value and cherish their watercourses and waterfront environments.

These three simple, but clear objectives have remained the same throughout the life of the Campaign.

The location of Mersey Basin Campaign

The Mersey Basin Campaign covers two river basins in the Northwest of England: the catchment of the River Mersey and the contiguous river catchment to the north, the River Ribble and its tributaries.

THE EMERGENCE OF THE PARTNERSHIP

The scale and complexity of the clean-up programme required to deal effectively with the gross water pollution and waterside dereliction was too great for any one authority or agency. (At the time, there was no national environmental programme for water quality improvements; that would not come until water privatisation in 1990).

"The task of cleaning up the Mersey – the watercourses and waterside areas of the whole catchment – is a comprehensive and formidable one. The task calls for a team effort, in which the inputs of all sectors encourage each other and generate a momentum of improvement greater than could be achieved otherwise." (Walton, 1982 p. 11)

This recognition that a combination of public, private and voluntary sector action was necessary to bring about the total process of renewal for the water and bordering land led to a new partnership being formed in 1985. At the time, most partnerships operated just between the public (government) and private (business) sectors. The Mersey Basin Campaign partnership was conceived differently from the start. It was organised around an independent Chair leading a unit from the government's Department of the Environment.

Key partners were brought in, including the (then publicly-owned) water authority, local government authority representatives and professional officers with an advisory role from a number of non-governmental organisations. Interestingly, the partnership had no industry involvement at the time, and it wasn't until 1987 when ICI, a multinational chemical company, joined as the first industrial partner, that the private sector joined in with the work of the Campaign.

Within the context of Mersey Basin Campaign, partnership is taken to mean an agreement between two or more partners to work together to achieve common aims.

GOVERNANCE, MANAGEMENT AND DECISION-MAKING

The organisational structure of the Mersey Basin Campaign

The organisation of Mersey Basin Campaign has changed over its life span to reflect regional changes and developments and to build upon experience. Two major changes in structure have taken place over its 19-year history to reflect the growing and changing needs of the Campaign and the region.

1. In 1996, Mersey Basin Campaign gained some independence from government as it became an'arms length management organisation'. This 'privatising' was considered to be necessary to allow the Campaign to be more effective for engagement of the private and voluntary sectors. Although now at arm's length, the Mersey Basin Campaign still retains its part sponsorship from the UK central government, and has enjoyed the ongoing support of successive governments since its inception in 1985.
2. A review of the partnership in 2001 brought about further changes to the governance of the Campaign, allowing even wider participation in the Campaign and its work through changes to the organisational structure, and the development of a constitution for the Campaign Council.

These two fundamental changes have led the Campaign to develop from a government-run initiative, led by an independent Chair, to its current partnership status discussed below.

The partnership approach

The partnership of the Mersey Basin Campaign in 2004 is based upon active involvement through a number of organisational structures (Figure 1).

This partnership governance can be summarised as:

- the Council, as the formally constituted, non-executive governing body of the Campaign, determines strategic guidance and approves the annual corporate plan;
- the Mersey Basin Business Foundation (MBBF) is a mechanism for business and financial management, contracts and the employment of the Mersey Basin Campaign staff. It is the legal personality of the Campaign;
- the Healthy Waterways Trust is the Campaign's charitable arm and is a registered environmental body; and,
- five Advisory Groups providing focus for more specific policy development and guidance for the work of the Campaign.

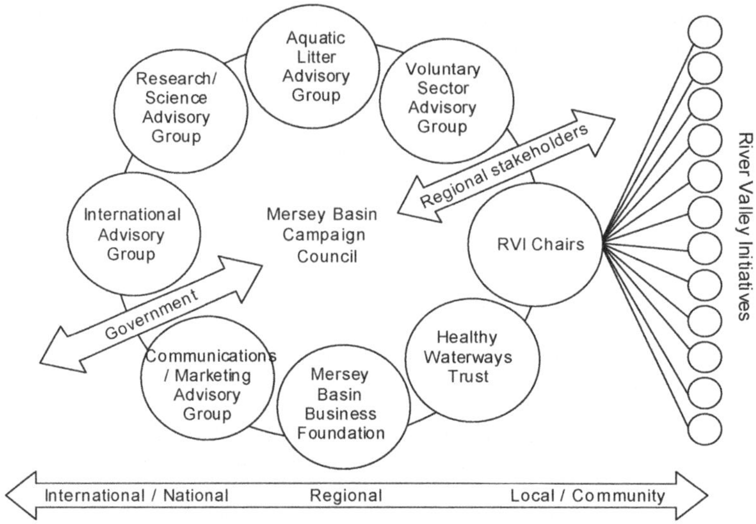

Figure 1. Organisational structure of the Mersey Basin Campaign

This partnership governance can be summarised as:

- the Council, as the formally constituted, non-executive governing body of the Campaign, determines strategic guidance and approves the annual corporate plan;
- the Mersey Basin Business Foundation (MBBF) is a mechanism for business and financial management, contracts and the employment of the Mersey Basin Campaign staff. It is the legal personality of the Campaign;
- the Healthy Waterways Trust is the Campaign's charitable arm and is a registered environmental body; and,
- five Advisory Groups providing focus for more specific policy development and guidance for the work of the Campaign.

Some partners play a key role within the Mersey Basin Campaign, sitting on the Campaign Council or as Directors of MBBF, whilst others perform more peripheral roles, acting as advisers on specific issues of the Campaign's work. The nature of the roles partners play within the Campaign's governance and management is detailed in section 3.3.

The Campaign structure also allows spatial flexibility in the partnerships. Regional stakeholders play key roles within the Campaign Council and MBBF. Local stakeholders partner the Campaign through its local River Valley Initiatives (see Wood, Handley & Kidd 1999; Kidd & Shaw, 2000; Salthouse, 2000). RVIs reflect local challenges and needs and are composed of key local partners, with the Campaign employing a number of river valley co-ordinators to support the individual steering groups and to take decisions forward through fund raising, managing projects, events and awareness raising activities.

The decision makers and the Campaign managers

The Council

The Council was set up as the governing body for the Mersey Basin Campaign within which key regional stakeholders provide strategic direction and policy guidance to the Campaign in delivering its objectives. It is an unincorporated stakeholder partnership of 38 representatives with two types of members: partners, with voting rights; and, advisers/observers without voting rights.

The Campaign sought views from across the region on the appropriate stakeholders for inclusion within the Council through a formal consultation, endorsed by government. Members to the Council are appointed as representatives for their organisations, sectors or area of interest. Core partners on the Council include representatives of the water company (United Utilities); the environmental regulator (Environment Agency); local government; the Regional (economic) Development Agency; and a number of other public bodies (for example, English Nature). The voluntary sector is also represented on the Council, with the Voluntary Sector Forum, and advisory group to the Campaign Council, providing representation. Importantly, the Council's constitution allows its membership to be dynamic, reflecting regional changes over the 25-year life of the Campaign.

The Mersey Basin Business Foundation

The Mersey Basin Business Foundation, a non-profit making limited company, carries out the task of overall operational management for the Campaign. Directors are partners from industry, based on an initial partnership between the Campaign, ICI (in 1987), Shell (1988) and Unilever (1989). MBBF was launched as a separate and increasingly important arm of the Campaign in 1992 and now has 12 members. It actively seeks to expand the number of businesses linked to the Campaign and specific Campaign projects. Member organisations are encouraged to incorporate Campaign objectives into their daily activities and business practices. The MBBF is the recipient of the core government grant to the Campaign.

These two structures (the Campaign Council and MBBF) enable partners to work with the Campaign at different levels and degrees of commitment. The MBC has found that one of the keys to successful partnerships is ensuring there are many opportunities to build relationships with partners and connections at all organisational levels, from the top of an organisation down. For long-term commitment, we have found it essential to establish a relationship at the top of the organisation, thereby ensuring commitment throughout that organisation (Kim 2002).

BUILDING TRUST AND DEVELOPING A LONG-TERM PARTNERSHIP

One potential stumbling block to good partnership working is lack of consistency between partners' objectives. Partnerships such as the Mersey Basin Campaign bring together disparate groups and sectors to work together towards a shared mission. The Campaign has its own unique objectives and mission. Partner organisations, especially core partners, generally have objectives which complement at least one of the 3 key Campaign objectives.

One of the strengths of the partnership has been the relationship between United Utilities Water plc and the Campaign. Prior to water privatisation, the North West Water Authority (NWWA) had been investing to deal, *inter alia*, with the crude waste discharges to the Mersey. Collaborating with the Campaign, industry matched NWWA's investment, cleaning up its own discharges.

On privatisation of water services in 1989, the then North West Water ltd (which later became part of United Utilities plc) was charged with delivering a £2.2 billion, five-year investment programme to improve the region's water infrastructure and make inroads into the worst of the region's water pollution. The company's objectives mirrored those of the Campaign, and for the first time, the necessary investment existed to make these objectives a reality. The company and Campaign have enjoyed a constructive, collaborative relationship ever since, across all levels of the organisations. For United Utilities, the Campaign provides a broader context for environmental improvements, helping make connections to waterside regeneration and providing third-party publicity of the United Utilities' water quality achievements. For the Campaign, United Utilities is a core funder, the main contributor to improving water quality, and a source of support and technical advice.

But how do partners collaborate if their objectives are not the same? By establishing and identifying the benefits in working together the complementary nature of the respective organisations objectives becomes more focused and allows the partners to recognise the added-value in working together. In the case of Mersey Basin Campaign, collaboration on a project basis over the years has allowed trust to develop between the partners through working together, producing positive results and both partners obtaining benefits from working together rather than individually.

For example, businesses are keen to work with organisations that can supply niches they do not supply themselves, for example delivering community engagement. These opportunities enable the partnership and private sector to develop shared vocabularies, finding overlaps and understanding. Such an approach enables the delivery of complex regeneration projects, which a single partner alone cannot realise.

Only once the benefits have been established can trust develop between the partners. The government origins and continued support for Mersey Basin Campaign has provided the basis from which a good reputation has been developed. This reputation has developed over time from building upon achievements in water quality improvements within the catchments and a having a good track record in project delivery. This has been achieved through working with the partners gaining continued improvements in water quality in the two catchments and large-scale regeneration programmes, especially targeted at derelict waterside environments.

IMPLEMENTING ACTION

How do we take forward decisions and put our actions into practice?

The Campaign is essentially a project-based organisation, with projects delivered at the catchment or sub-catchment scale. The decision makers at the catchment scale are the Council members. However, at the sub-catchment scale that decision making transfers to an individual steering group of the Campaign's River Valley Initiatives (RVI), where partners play a key role through the steering groups for each RVI, or through specific projects and action within the RVIs work. For example, a partner may provide the Chair for a RVI steering group, leading the local initiative. Alternatively, a partner may participate within a specific project being undertaken by a RVI co-ordinator, such as a clean-up event or habitat rehabilitation programme. The Campaign has found this action-led partnership approach has been successful in maintaining interest in the Campaign and its work by partners. This organisational structure provides a framework in which all partners can gain from their own inputs.

CHALLENGES OF WORKING IN PARTNERSHIPS

Measuring success: what are the lessons learnt from partnership working?

There are a number of benefits and difficulties from partnership working. Tippett (2001) discusses these in detail, however, they are summarised within this paper. Whilst the benefits can be visually obvious seen through the improvements to water quality within the catchments and the enhanced waterside environments currently undergoing regeneration, such as can be seen at The Quays in Salford, the difficulties tend to be less visible, but nonetheless as important. These benefits and difficulties are summarised below (see Tippett, 2001 for further details).

The benefits include:

- Partnerships produce results beyond the capabilities of the individual entities;

- It is a cost effective way of working;
- Partnerships can provide an increased knowledge base for strategic planning;
- They enhance the possibilities for integrated thinking through a combination of multiple perspectives, especially those involving all sectors;
- A partnership can ensure a broad range of support for projects;
- They can be effective mechanisms for bringing about behavioural changes, certainly within the participating organisations;
- The participants to the partnership are more easily able to view how particular actions fit into the bigger picture;
- There is the possibility of increased learning amongst the participants.

In comparison to the benefits, there are far fewer disadvantages. Partnership working can be perceived as creating "additional" work. Within a partnership responsibilities for specific work can be shifted or passed to another partner, thereby increasing the workload for partners.

There are also a number of difficulties that can be recognised in setting up and maintaining a partnership. These include:

- It takes time to develop a working relationship within the partnership;
- Continuous effort and attention are required in order for the partnership to remain viable;
- It is very difficult to maintain enthusiasm over a long period of time, especially with those connected only tangentially to the initiative;
- Both the partnership and the external context change over time;
- The timetable of development needs to be driven by participants' needs rather than by the often rigid requirements of funding deadlines;
- It is often difficult to measure and cost the added value brought by the partnership;
- The benefits tend to accrue over the longer term rather than immediately;
- There is a need to credit partners amongst whom the responsibilities are shared.

These difficulties can be overcome by ensuring effective management and ownership of the partnership by the partners. A clear vision, which is understood and shared by all the partners, with broad and ambitious goals is needed thereby ensuring expectations are managed, and preferably, exceeded.

What are the critical success factors?

There are a number of factors that have contributed to the successful long-term partnership that is the Mersey Basin Campaign, that have allowed the improvements within the region, to the extent that in 1999, Mersey Basin Campaign won the inaugural World River Prize. These include:

- The partnership approach – the Campaign is a partnership that includes government, agencies (such as the Environment Agency), local authorities, businesses – corporate bodies such as United Utilities and, of course, local communities;
- A clear vision from inception with an undiluted focus on water, watersides and communities;
- A realistic timescale for change to occur; the Campaign was set up as a 25 year initiative;
- Continued government support – successive governments and individual ministers have strongly and actively backed the Mersey Basin Campaign;
- Strategic influencing followed through with delivery - the Campaign operates at the global, EU, national and regional level but also through 19 locally driven, locally focused river valley initiatives working at the grassroots level;

- An evidenced based organisation, with strong scientific and technological innovation such as the oxygenation of the Manchester Ship Canal at The Quays at Salford;
- Resources – the current five year Asset Management Programme of the water company, United Utilities, exceeds £2 billion for 2000 – 2005 investment;
- Leadership – the succession of government appointed, independent chairs of the Campaign leading the partnership with determination and drive.

CONCLUSION

The Mersey Basin Campaign was set-up by the British government in 1985 as a partnership to tackle the legacy of Industrialisation in Northwest England. This partnership has changed over the life span of the Campaign, building from its government roots to reflect the complexity of the twin problems of poor water quality and industrial dereliction. The partnership recognises the interdependence of economic prosperity and quality of the environment and the need for a sustainable development approach. This required a combination of public, private and voluntary sector action within the partnership. The current governance structure of the Campaign provides a number of opportunities for partners to work with the Campaign at different levels and degrees of commitment. MBC have found that one of the keys to successful partnerships is ensuring there are many opportunities to build relationships with partners and connections at all organisational levels.

Although there are clear benefits from partnership working, especially amongst those partners with strong similarities in organisational objectives, there are also some disadvantages and difficulties. Partnerships require time and effort in setting up and maintaining the relationship. However, a clear vision, which is understood and shared by all partners, with broad and ambitious goals can allow expectations to be managed and exceeded and the partnership to strengthen.

REFERENCES

H.M.S.O. (1874) Royal Commission on Rivers Pollution: 1868-74, Final report. H.M.S.O. London.

Kidd, S & Shaw, D. (2000) The Mersey Basin and its River Valley Initiatives: an appropriate model for the management of rivers? Local Environment, **5**(2) 191-209.

Kim, J.S. (2002) A collaborative partnership approach to integrated waterside revitalisation: the experience of the Mersey Bain Campaign, North West England. PhD thesis, Dept. of Civic Design, University of Liverpool.

MBC (1995) Mersey Basin Campaign, First Periodic Review: Changing for the better. Mersey Basin Campaign, Manchester, UK.

Salthouse, C. (2000) Making the most of the Mersey Estuary: a partnership approach to catchment management. International Journal of Urban Sciences, **4**(2), 129-138.

Tippett, J. (2001). Integrated Catchment Management and Planning for Sustainability - the Case of the Mersey Basin Campaign. Dissertation submitted for MA (Econ.) in Social Research Methods. Faculty of Social Sciences and Law, University of Manchester.

Walton, P. (1982). Cleaning up the Mersey: a consultation paper on tackling water pollution in the rivers and canals of the Mersey catchment, and improving the appearance and use of their banks. London, Department of the Environment, UK.

Watson, N.M. (2001) The Contribution of Multi-Stakeholder Partnerships to Sustainable River Basin Development in Canada. Proceedings of the 7th International Sustainable Development Research Conference, University of Manchester, UK, April 5-6, pp.319-326.

Wood, R., Handley, J & Kidd, S. (1999) Sustainable development and institutional design: the example of the Mersey Basin Campaign. Journal of Environmental Planning and Management, **42**(3), 341-354.

Development of a Danube River Basin Management Plan in line with the requirements of the EU Water Framework Directive

Ursula Schmedtje*, Philip E. Weller*

International Commission for the Protection of the Danube River, Permanent Secretariat, Vienna International Center D0412, P.O. Box 500, 1400 Vienna, AUSTRIA (Email: icpdr@unvienna.org)

Abstract The Danube River Basin is the second largest river basin in Europe and the most international one with 18 countries having territories in the basin. The EU Water Framework Directive requests the Member State or Member States concerned to endeavour to establish appropriate coordination with the relevant non-Member States. The International Commission for the Protection of the Danube River (ICPDR) serves as the platform for coordination to develop and establish the Danube River Basin Management Plan (DRBMP). This paper describes the coordination mechanisms for WFD implementation as well as the development of the Danube River Basin Management Plan agreed in the frame of ICPDR.

Keywords Danube river basin, EU Water Framework Directive, ICPDR, transboundary cooperation, river basin management plan

Introduction

On December 22, 2000 the Water Framework Directive 2000/60/EC (WFD) came into force. The EU Member States are obliged to fulfil this Directive. The WFD brings major changes in water management practices. Most importantly, it:

- sets uniform standards in water policy throughout the European Union and integrates different policy areas involving water issues,
- introduces the river basin approach for the development of integrated and coordinated river basin management for all European river systems,
- stipulates a defined time-frame for the achievement of the good status of surface water and groundwater,
- introduces the economic analysis of water use in order to estimate the most cost-effective combination of measures in respect to water uses,
- includes public participation in the development of river basin management plans encouraging active involvement of interested parties including stakeholders, non-governmental organisations and citizens.

What makes the implementation process in the Danube River Basin a particular challenge is the fact that only some countries are EU Members and therefore obliged to fulfil the WFD. Besides Austria,Germany and Italy, five additional Danube countries have become EU Members States on May 1, 2004. Three other Danube countries are in the process of accession and are preparing to fulfil the complete body of EU legislation in order to become EU Members. Others have not initiated a formal process to join the EU.

 ISBN: 184339510X. Published by IWA Publishing, London, UK.

The Danube River Basin falls into the third category of rivers addressed in Art. 13.3 WFD – an international river basin extending beyond the boundaries of the European Community. The WFD requests the Member State or Member States concerned to *„endeavour to establish appropriate coordination with the relevant non-Member States, with the aim of achieving the objectives of this Directive throughout the river basin district."* (Art. 3.5 WFD). Therefore, appropriate coordination mechanisms need to be established. The International Commission for the Protection of the Danube River (ICPDR) has been utilised to secure this coordination.

The Danube River Basin

The Danube River Basin is the second largest river basin of Europe, covering 801 463 km². It is also the most international river basin including territories of 18 countries (see Table 1).

Table 1 Coverage of the states in the Danube River Basin (DRB) and estimated population (data source: Competent authorities in the DRB unless marked otherwise; MS = EU Member State, CC = Candidate Country)

State	Code	Status in European Union	Coverage in DRB [km²][1]	Percentage of DRB [%]	Percentage of DRB in state [%]	Population in DRB [Mio.]	Percent of population in DRB [%]
Albania	AL	-	126	< 0.1	0.01	< 0.01	< 0.01
Austria	AT	MS	80 423	10.0	96.1	7.7	9.51
Bosnia-Herzegovina	BA	-	36 636	4.6	74.9	2.9	3.58
Bulgaria	BG	CC	47 413	5.9	43.0	3.5	4.32
Croatia	HR	CC	34 965	4.4	62.5	3.0	3.71
Czech Republic	CZ	MS	22 870	2.9	31.1	2.8	3.46
Germany	DE	MS	56 184	7.0	16.8	9.3	11.49
Hungary	HU	MS	92 778	11.6	100.0	10.2	12.60
Italy [2]	IT	MS	560	< 0.1	0.2	0.02	0.02
Macedonia	MK	-	109	< 0.1	0.2	< 0.01	< 0.01
Moldova	MD	-	12 834	1.6	35.6	1.1	1.36
Poland	PL	MS	430	< 0.1	0.1	0.04	0.05
Romania	RO	CC	231 511	29.0	97.4	21.0	25.94
Serbia-Montenegro	CS	-	88 635	11.1	90.0	9.8	12.11
Slovak Republic	SK	MS	47 271	5.9	96.0	5.2	6.42
Slovenia	SI	MS	16 389	2.0	81.0	1.7	2.10
Switzerland	CH	-	1 809	0.2	4.3	0.02	0.02
Ukraine	UA	-	30 520	3.8	5.4	2.65	3.27
Total			801 463	100		80.95	100

[1] For the purpose of comparison the coverage of the states was calculated using GIS based on the DRBD overview map. These values differ slightly from the official data of some countries, since other methods of calculation have been used.

[2] Data source: Autonomous Province of Bozen – South Tyrol.

Some of the Danube sub-basins have the size of other important international river basins in Europe. The Tisza River Basin is the largest sub-basin in the Danube River Basin (157 186 km²) and is a bit larger than the Elbe River basin (148,268 km²). The Tisza River Basin covers the territory of 5 countries (Ukraine, Romania, Hungary, Slovakia and Serbia-Montenegro). The Sava River is the largest Danube tributary by discharge (average 1564 m³/sec) and the second largest by catchment area (95,419 km²). The Sava watershed is almost equivalent to the Odra River Basin (118,000 km²) although its discharge is almost twice as large (877.3 m³/sec). The Sava River Basin covers in part territories of Slovenia, Croatia, Bosnia-Herzegovina and Serbia-Montenegro.

The Danube River Basin, however, is not only impressive because of its size. It contains highly diverse geographic regions, and has a rich human history. It is also characterised by major socio-economic differences among the countries with a gradient from upstream to downstream. This fact greatly influences the basin wide management of the water resources, restoration and pollution reduction efforts. For example, the percentage of population connected to public water supply as well as to sewage disposal shows great differences and standards from West to East.

As requested in Art. 3.1 WFD *"Member States shall identify the individual river basins lying within their national territory ... and shall assign them to individual river basin districts."* Under the framework of the ICPDR the Danube River Basin District has been established covering 1) the Danube River Basin, 2) the Black Sea coastal catchments on Romanian territory, and 3) the Black Sea coastal waters along the Romanian and partly the Ukrainian coast (see Table 2).

Table 2 Area of the Danube River Basin District

	Territory	Official area [km²]	Digitally determined area [km²][1]
Danube River Basin (DRB)	18 countries (see Table 1)		801 463
Black Sea coastal river basins	Romania	5 198	5 122
Black Sea coastal waters	Romania and Ukraine		1 242
Danube River Basin District (DRBD)			807 827

[1] For the purpose of comparison the areas were calculated using GIS on the basis of the DRBD overview map. The value for the Black Sea coastal river basins differs slightly from the official data, since other methods of calculation have been used.

Figure 1 shows the geographical coverage of the Danube River Basin District. The outer boundary of the Danube River Basin District was defined taking into consideration the hydrological boundaries of the surface waters and groundwater. In addition to the Danube River Basin, the small coastal basins of the Black Sea tributaries lying on Romanian territory between the eastern boundary of the DRB and the coastal waters of the Black Sea have been included in the Danube River Basin District. These coastal catchments were included in the DRBD, because they influence the coastal waters along the Romanian coastline. The coastal waters of the Black Sea play an important role in the pressure and impact assessment, since the nutrient flow from the Danube River Basin contributes significantly to nutrient concentrations in the Black Sea and are therefore a key cause for the eutrophication of the sea.

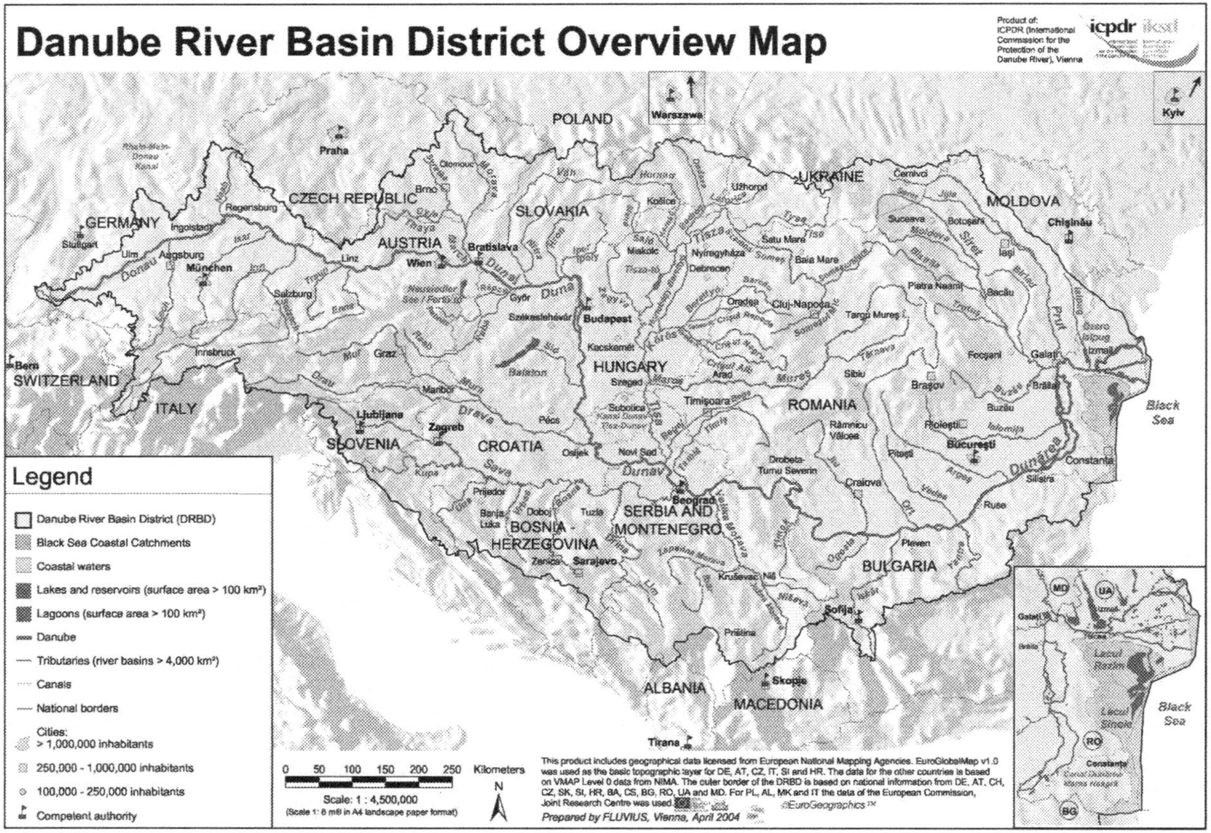

Figure 1 Geographical coverage of the Danube River Basin District

Small sections of the Danube river are still rather untouched ecosystems and, despite possible pollution problems, a unique heritage to be preserved. In addition, the Danube River Basin still hosts many species and habitats of outstanding ecological value and unique importance for biodiversity such as wetlands and nature protection sites. In particular, the Danube Delta is of global significance. The future management of the river basin needs to ensure that the focus of measures is not only the restoration of affected water bodies but equally important the preservation of those areas where the status is still high.

In recent decades, considerable improvements in environmental conditions in the Danube basin have been made. Where investments, e.g. in wastewater treatment, have taken place, the improvement of the water quality is clearly visible. However, a major part of pollution reduction can be attributed to the decline of industries and agricultural activities in the middle and lower part of the basin. The main pressures are due to organic pollution, hazardous substances and nutrients. In addition, hydromorphological alterations are the cause of degradation throughout the basin. In these areas, investments for a sustainable reduction of pollution levels as well as from other pressures has just started and will have to continue for another 10 to 20 years.

International relationships to ensure coordination

"*In the case of an international river basin district extending beyond the boundaries of the Community, Member States shall endeavour to produce a single river basin management plan, and, where this is not possible, the plan shall at least cover the portion of the international river basin district lying within the territory of the Member State concerned.*" (Art. 13.3 WFD).

The "Convention on Cooperation for the Protection and Sustainable Use of the Danube River" (Danube River Protection Convention, DRPC) forms the overall legal instrument for cooperation and transboundary water management in the Danube River Basin. The main objective of the convention is the sustainable and equitable use of surface waters and groundwater and includes the conservation and restoration of ecosystems. The Contracting Parties cooperate on fundamental water management issues and take all appropriate legal, administrative and technical measures, to maintain and improve the quality of the Danube River and its environment. Presently, Austria, Bulgaria, Croatia, the Czech Republic, Germany, Hungary, Moldova, Romania, the Slovak Republic, Slovenia, Serbia and Montenegro, Ukraine and the European Community are Contracting Parties to the DRPC. Bosnia and Herzegovina has an observer status. To facilitate the implementation of the DRPC, the Danubian states agreed that with its entry into force the ICPDR is established.

In November 2000 all Contracting Parties of the Danube River Protection Convention stated their commitment to implement the WFD within their jurisdiction and to cooperate in the framework of the ICPDR to achieve a single, basin-wide coordinated Danube River Basin Management Plan. For states with territories of less than 2 000 km² in the Danube River Basin (DRB) the ICPDR has attempted to establish appropriate bilateral coordination.

In view of the size and number of states that have territories in the Danube basin coordination is required on different levels in order to fulfil *"the environmental objectives established under Article 4, and in particular all programmes of measures"* (Art. 3.4 WFD). For issues of basin-wide importance the ICPDR serves as the platform for coordination in the implementation of the WFD in the Danube River Basin District. Transboundary issues not covered by the ICPDR are solved at the appropriate level of cooperation e.g. in the frame of bilateral/multilateral river commissions. Local issues remain a national task. Generally, coordination will take place at the lowest level possible so that the coordination via the ICPDR can be limited to those issues necessary on the basin-wide level.

At its 3rd Ordinary Meeting on November 27-28, 2000 in Sofia the ICPDR adopted the following resolutions:

- The ICPDR will provide the platform for the coordination necessary to develop and establish the River Basin Management Plan for the Danube River Basin.
- The Contracting Parties ensure to make all efforts to arrive at a coordinated international River Basin Management Plan for the Danube River Basin.

In the ICPDR all Contracting Parties and Bosnia-Herzegovina as an observer support the implementation of the WFD in their territories and cooperate in the framework of the ICPDR to achieve a single, basin-wide coordinated Danube River Basin Management Plan. The ICPDR President has addressed the other DRB countries not cooperating under the DRPC to commit themselves to cooperate with the ICPDR to achieve a basin-wide coordinated DRBMP. Poland, Switzerland, Macedonia and Albania have offered their support. From Italy no response was received. On the operational level, it is the obligation of the Contracting Parties to ensure the necessary coordination with their DRB neighbours not cooperating under the DRPC.

The River Basin Management Expert Group was created to prepare and coordinate the necessary activities for the implementation of the WFD. All states cooperating under the DRPC are represented in the River Basin Management Expert Group. The group jointly agrees on the necessary actions for the development of the Danube River Basin Management Plan, e.g. the development of a strategy for establishing the RBM Plan, development of the roof report to the European Commission or identifying needs for harmonisation of methods and mechanisms (see Figure 2).

The Danubian states cooperating under the DRPC report regularly to the ICPDR on the progress of WFD implementation in their own states. These national reports serve as a means for exchanging information between the states and for streamlining the implementation activities on the national level. At each of its Ordinary Meetings and Standing Working Group Meetings the ICPDR deals with the step-wise implementation of the WFD in the Danube River Basin and takes the necessary decisions.

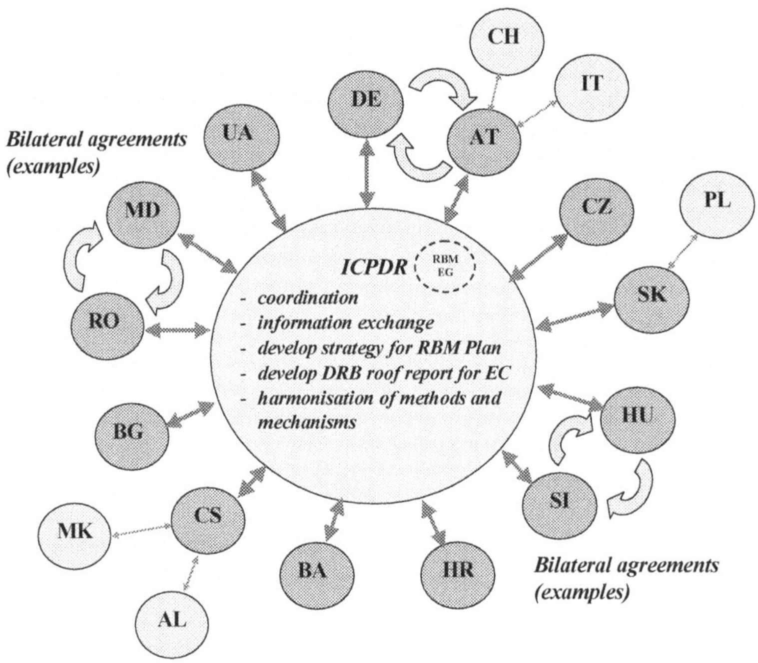

Figure 2 Coordination mechanisms for WFD implementation in the Danube River Basin

Development of The Danube River Basin Management Plan

Due to the large number of states and the coordination requirements in the Danube River Basin District (DRBD) it is necessary to divide the Danube River Basin Management Plan into two parts. Part A (roof of the DRBMP) gives relevant information of multilateral or basin-wide importance, whereas Part B (national input to DRBMP) gives all relevant further information on the national level as well as information coordinated on the bilateral level (see Figure 3). The ICPDR has a coordinating and supporting function, but does not report on its own. Each state will deliver the roof report (Part A) and its own national report (Part B).

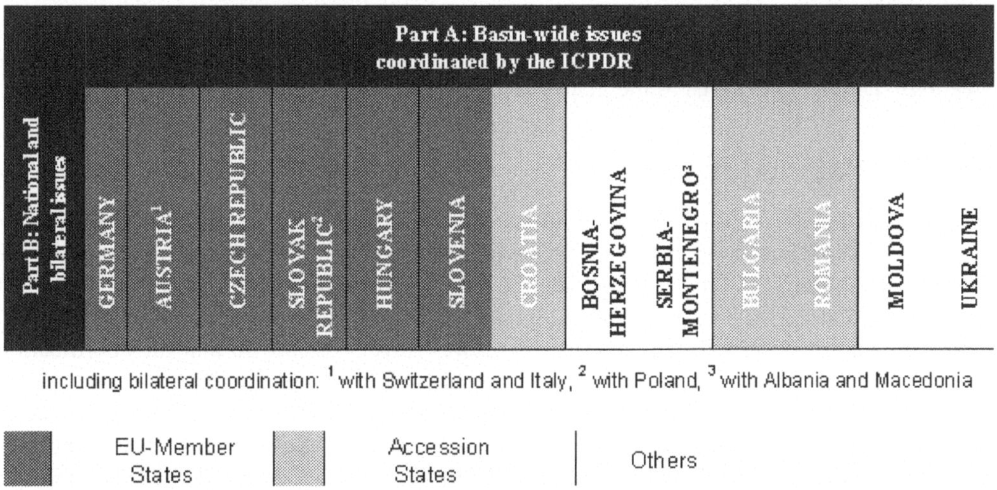

Figure 3: Structure of the Danube River Basin Management Plan. (This figure reflects the situation at the time of reporting (June 2004)).

Part A – Roof report

The Roof report contains information on issues of multilateral or basin-wide importance and demonstrates the basin-wide coordination arrangements. This includes, in particular, an overview of the main driving forces of multilateral or basin-wide importance and the related pressures exerted on the environment. The analysis is based on data delivered by the competent authorities in the Danube River Basin. Only where data is not available other data sources are used. The overview includes effects on coastal waters of the Black Sea as far as they are part of the Danube River Basin District. The Roof report intends to give an overview of the situation in the Danube river basin district as a whole and to set the frame for the understanding of the national reports.

A summary of the relevant information on surface waters and groundwater will be given in Part A. Detailed information will be in Parts B. The Roof report deals with the following surface waters (as shown in Figure 1):

- all rivers with a catchment size of > 4 000 km²,
- all lakes, reservoirs and lagoons with an area of at least 100 km², and
- the main canals.

Groundwaters are generally of local or regional importance and will be dealt with in the national reports. The Roof report deals only with the important transboundary groundwater bodies, in particular:

- all transboundary groundwater bodies > 4000 km², and
- transboundary groundwater bodies < 4000 km², if they are very important; the importance has to be based on certain bilaterally agreed criteria e.g. socio-economic importance, groundwater use, impacts, pressures, interaction with aquatic eco-systems.

The Roof report will be sent by the states to the European Commission with the National reports. The Roof report is identical for all states. The content of the management plan results from the work of the ICPDR expert groups and is approved by the ICPDR Ordinary Meetings. The ICPDR serves as the platform for coordination.

Part B – National reports

The national reports give all relevant further information on the national level as well as information coordinated on the bilateral level. Transboundary issues not covered by the ICPDR are solved at the appropriate level of cooperation e.g. in the frame of bilateral/multilateral river commissions. Issues with local effects remain a national task. Generally, coordination will take place at the lowest level possible so that the coordination by the ICPDR can be limited to the basin-wide issues.

The 2-level outline of the Danube River Basin Report has been applied for the first time in the report prepared for the European Commission on the identification of the competent authorities in the Danube River Basin and the definition of the Danube River Basin District according to Art. 3.8 and Annex I WFD (Reporting deadline: June 22, 2004). For the report on the characterisation and analysis of the Danube River Basin in line with Art. 5 and Annex II and III WFD (Reporting deadline: 22 March, 2004) the 2-level outline will also be applied. The Danube River Basin Management Plan, which is due by the end of 2009, will most probably follow a 3-level approach, since it will be necessary to develop separate plans for the sub-basins in the Danube River Basin. A RBM Plan is currently under preparation for the Sava River Basin (shared by Slovenia, Croatia, Bosnia-Herzegovina and Serbia-Montenegro) as well as several sub-basins of the Tisza River Basin, namely the Somes/Szamos River Basin and the Crisul/Körös River Basin (both shared by Hungary and Romania) as well as for the Hornád/Hernád River Basin (shared by Hungary and Slovakia).

Conclusions

The countries of the Danube Region have been cooperating under the framework of the International Commission for the Protection of the Danube River to prepare a joint basin management plan. The first major step in that work – the characterisation of the basin – has now been almost completed and forms the basis for identifying the problems and additional efforts and actions needed to reduce pollution, and minimize other pressures negatively influencing the quality of water in the basin.

The cooperation in the Danube River Basin has progressed remarkably in the last 10 years since the signing of the Danube River Protection Convention. An excellent cooperation spirit has evolved. The EU Water Framework Directive has been a strong tool for setting the agenda in the Danube River Basin and for uniting efforts to achieve common goals. On a technical level, the work has strongly benefited by the support of the UNDP/GEF Danube

Regional Project as well as the chairmanship of the River Basin Management Group by the European Commission. Finally, the decision to use English as a common language (a foreign language for all) has greatly eased the communication and helped establish common ground for cooperation.

Acknowledgements

The strategy for the development of the Danube River Basin Management Plan has been developed under the guidance of the River Basin Management Expert Group (RBM EG) of the ICPDR. Our special thanks are due to

- the members (Austria, Bosnia-Herzegovina, Bulgaria, Croatia, Czech Republic, Germany, Hungary, Moldova, Romania, Serbia-Montenegro, Slovak Republic, Slovenia and Ukraine) and observers (WWF, GWP, IAD, DEF) of the RBM EG,
- the present and former Chairman of the RBM EG, Joachim D'Eugenio and Friedrich Barth,
- the present and former President of the ICPDR, Catherine Day and Fritz Holzwarth,
- the UNDP/GEF Danube Regional Project, Ivan Zavadsky and William A. Garner.

Restoration of the Hawkesbury-Nepean river system, New South Wales Australia

J. M. Anderson*

* Department of Commerce, McKell Building L13, 2-24 Rawson Place, Sydney NSW 2000, Australia (E-mail: john.anderson@commerce.nsw.gov.au)

Abstract There has been a significant decline in the health of Australian rivers in the 200 years since European settlement of Australia. The Hawkesbury-Nepean river system is the principal water source for the city of Sydney, capital of the state of New South Wales. The health of the river system has been degraded as a result of water extractions, urban and agricultural drainage and discharges of treated urban wastewater. Over the last ten years, the New South Wales state government has embarked on range of policy initiatives to restore the health of rivers in New South Wales. The paper discusses restoration measures for the Hawkesbury-Nepean river system including a newly announced Metropolitan Water Plan.

Keywords river basin; restoration; Hawkesbury-Nepean

Introduction

Australia was first settled by Europeans 215 years ago. Prior to European settlement the small indigenous population lived in harmony with the landscape and the rivers. The state of Australia's rivers at the time of European settlement can be judged by the following extracts from the diaries of the explorer, Captain Charles Sturt who explored the lower reaches of Australia's longest river, the Murray, in late 1829. This is the scene described by Sturt when he first sailed onto the Murray at its junction with the Murrumbidgee in south-western New South Wales (Dunmore-Lang 1834).

"...a river of noble breadth, about 350 feet wide and from 12 to 20 feet in depth.
"Its reaches were from half to three-quarters of a mile in length, and the views on it were splendid:
its transparent waters were running over a sandy bed at two and a half knots an hour..."

In the NSW Art Gallery there is an 1896 painting of the Hawkesbury river by the Australian impressionist, Arthur Streeton, entitled 'Purple Noon's Transparent Might'. Again that word 'transparent' tells us about the state of the rivers just over 100 years ago.

Since then the rivers of New South Wales have been greatly altered by the construction of storages in the headwaters to provide regulated flow for irrigation and urban development. In the Murray the natural flow patterns of high winter-spring flows and low summer-autumn flows have been entirely reversed by irrigation storage and regulated flow releases. Flows in the lower reaches at the place of Sturt's 1829 description are greatly diminished due to upstream diversions, the currents are sluggish and the waters are no longer transparent. Sturt's words tell us starkly what it is that we have lost.

Hawkesbury-Nepean River

The Hawkesbury-Nepean river basin has a catchment area of 22,000km^2, an average annual rainfall of 890mm per year, and an average annual discharge of about 3,000Mm^3 per year. Like all Australian catchments, the rainfall and the streamflow is highly variable. The Hawkesbury enters the Pacific Ocean about 30km north of Sydney on the east coast of Australia.

The Hawkesbury-Nepean river system is the principal water source for the city of Sydney, capital of the state of New South Wales. Since the Nepean river was first tapped for Sydney's water supply in 1886, the population of Sydney has grown from less than 1 million to more than 4 million. The population is expected to grow by another million over the next 20 years.

Sydney Water Supply System

The major Sydney water storages command about half of the Hawkesbury-Nepean catchment and have average annual inflows of about 1,600Mm^3. In addition there is a diversion system to pump water from the Shoalhaven river, 160km south of Sydney.

The current Sydney water supply system has a yield of about 600Mm^3 per year. Water consumption in 2002-2003 reached 635Mm^3. New South Wales has experienced a sequence of severe droughts since 1990 and storage levels are currently below 45%. One newspaper recently described Sydney as: "Pools paradise, running on empty".

Loved Almost to Ruin

In the 20th century the Hawkesbury-Nepean river system has been loved almost to ruin. The catalogue of 20th century excesses include:

- Land use changes and urbanisation
- Storages, diversions, reduced flows
- Effluent and stormwater discharges
- Pollution and rubbish dumping
- Sand and gravel extraction
- Loss of riparian vegetation
- Decline in water quality
- Aquatic weed growth
- Invasion of exotic plants and animals
- Loss of biodiversity.

The impacts are most clearly manifested in declining water quality with the growth of aquatic weeds and some blue-green algae outbreaks in the lower reaches.

Water Reforms

Since first elected in 1995,the current New South Wales state government has embarked on a series of major water reforms.

Health Rivers Commission

In 1995 the government appointed a Healthy Rivers Commission to conduct inquiries into the health of New South Wales rivers and to make recommendations for improvements. The government published interim River Flow and Quality Objectives for rivers in 1997. The commission conducted an Inquiry into the Hawkesbury-Nepean system in 1998 and published extensive recommendations to improve river health.

Sydney Catchment Authority

Following a major water quality incident in the Sydney water supply system in 1998, the government appointed a Sydney Catchment Authority in 1999 with a charter to improve the protection of the water catchments. It also introduced new planning controls to manage development in the catchments which might impact on water quality.

Water Management Act

In 2000 the government replaced the Water Act 1912 with a new NSW Water Management Act 2000. Key elements of the new Act are:

- Water management for sustainable use
- New water access framework to prevent over-allocation
- Water sharing plans to share available water between users and environment.

Reform of the Water Agencies

The state's water agencies have been reformed with the creation of a Department of Infrastructure, Planning and Natural Resources in 2003. The new department has carriage of water resource management, the water reforms and the implementation of water sharing plans. Catchment management authorities have been created for each of the 35 major river basins in New South Wales.

Hawkesbury-Nepean River Management Forum

In 2001 the New South Wales government appointed a Hawkesbury-Nepean River Management Forum. The Forum's terms of reference were:

- To recommend environmental flows
- To advise on integrated water management for Sydney.

After extensive scientific studies, the Forum reported back to the government in April 2004. The Forum has recommended interim minimum environmental flows below the Nepean storages meeting a 20% transparency requirement and a 20% translucency requirement. This allocates all inflows upto the 80th percentile flow plus a minimum of 20% of flows in excess of the 80th percentile flow as environmental flows. All inflows to the Warragamba storage upto the 95th percentile flow plus a minimum of 20% of flows in excess of the 95th percentile flow will be allocated as environmental flows. In addition some specific contingency flows will be allowed for:

- Warragamba 3.0Mm^3/yr for aquatic weed control
- Upper Nepean 5.6Mm^3/yr for flushing/scouring and 2.7Mm^3/yr for fish spawning.

The Forum has also made recommendations for the provision of fishways and improved passage of environmental flows a number of existing weirs on the Nepean river. The Forum also suggested the setting of demand targets for Sydney of 570Mm^3/yr in 2008 and 550Mm^3/yr in 2011, and modification of system reliability from 97% to 95%

The environmental flows are expected to improve water quality considerably and increase the length of river that is swimmable by 40%. The effect of the interim flows will be monitored with a view to deciding on final environmental flow regime for the Hawkesbury river by 2015.

BASIX

The government has introduced a new Building Sustainability Index, known by the acronym BASIX (DIPNR 2004a). Under the BASIX provisions, all new houses in Sydney are required to achieve a 40% reduction in water use compared to the current Sydney average as a condition of development approval from 1 July 2004. The BASIX development approval scheme will be extended to all other local government areas in New South Wales from 1 July 2005.

The BASIX requirements effectively make the installation of water efficient appliances and fittings mandatory. Effectively, they also strongly encourage rainwater collection or recycled water systems to meet garden water needs. Further details of the BASIX scheme can be found on the DIPNR website.

Metropolitan Water Plan

A new Sydney Metropolitan Water Plan adopted by New South Wales Cabinet in September 2004. Details were announced during National Water Week in October 2004 (DIPNR 2004b). The proposed measures are expected to free up sufficient water to meet environmental flow requirements in the Hawkesbury-Nepean system and to provide for growth in demands for the next 30 years.

Demand Side Measures

Water Efficiency

Major commercial, industrial, government and local government users will be required to prepare water efficiency plans by March 2006 and to implement water efficiency measures by September 2007. The government is making available a Demand Management Fund to support capital investment in water efficiency initiatives.

The government has also announced an extension of the BASIX water efficiency program such that from July 2007 existing properties will require certification of compliance with water efficiency requirements when placed on the market for sale.

In 2005 the government will require manufacturers and retailers to display a label showing the level of water efficiency on appliances such as toilets, showerheads, clothes washers and dishwashers. The government and industry associations will also introduce a 'SmartWater Mark' labelling scheme for garden appliances and garden plants. Toilets that are not water-efficient will be removed from sale in 2005.

The government will continue to support householders who want to install water-efficient devices in their homes. The current subsidised household water audit program will continue with householders making a $22 contribution to the cost of the water audit and retrofitting of water efficient fittings. The current rainwater tank rebate program will be extended until July 2008.

Water Education

Building on the success of current water reduction programs, the government will implement a new water education program for New South Wales to maintain and increase community awareness of the need for water conservation. The 'Water Wise on the Farm' program will continue to improve agricultural water efficiency.

Pricing

Sydney currently has a two-part tariff with a fixed charge and a volumetric charge for all water used. The government is currently considering recommendations from the Independent Pricing and Regulatory Tribunal that outline a new tariff structure where households will pay a higher price for water use above a reasonable volume.

Leakage Reduction

Sydney Water will undertake a further extensive program of main renewal and leakage reduction to reduce system losses.

Supply Side Measures

The Sydney Catchment Authority is planning a number of measures to improve the capacity and security of the existing water supply headworks.

Existing Water Supply System

It is proposed to provide access to the deep storage below the current minimum operating levels in the water supply storages to provide a small increase in drought security (about 30Mm^3/year yield).

It is proposed to alter the current pumping rules for transfer of water from the Shoalhaven river to Sydney to increase the amount of water transferred during wet periods. Consideration is also being given to some increases in the capacity of parts of the transfer system. There is potential to increase system yield by between 60Mm^3 per year and 110Mm^3 per year, depending on the extent of works undertaken.

Water Recycling

Sydney currently reuses about 14Mm^3 per year and expects to this to 22Mm^3 per year in the near future through planned projects. Current projects include reuse for major industrial needs and residential reuse for garden watering and toilet flushing in new development areas.

The government will also undertake detailed planning for construction of a major water recycling scheme in western Sydney. The scheme will use water from the western treatment plants which discharge to the Hawkesbury-Nepean to supply recycled water to new greenfield residential development areas (24Mm^3 per year), agricultural uses (30Mm^3 per year), and with appropriate nutrient removal, environmental flows below the storages (upto 40Mm^3 per year). As well as reducing the demand on drinking water supplies, this proposal has the potential to reduce stress on urban streams by capturing treatment plant discharges. This recycling proposal is seen as a key initiative to restore the health of the Hawkesbury-Nepean river system and its tributary streams.

Groundwater

There are some groundwater aquifers within reasonable distance of Sydney. Further investigations will be undertaken as a drought contingency measure.

Desalination

Sydney Water will undertake a detailed evaluation of desalination as a contingency option to ensure that, if the current drought continues beyond another two years, a desalination plant could be constructed quickly if needed.

Conclusions

We cannot hope to turn back the clock and restore our rivers to 'transparent might' and the pristine conditions which existed prior to European settlement of Australia. There is, however, a great deal we can do to improve the efficiency of our water use and to restore sufficient environmental allocations to ensure improvements in river water quality.

The water policies, water reforms and recent implementation measures announced by the New South Wales state government represent a major commitment towards the restoration of the Hawkesbury-Nepean river system.

References

DIPNR (2004a). *BASIX – Building Sustainability Index*. NSW Department of Infrastructure, Planning and Natural Resources, May 2004, www.basix.nsw.gov.au.

DIPNR (2004b). *Meeting the challenges – Securing Sydney's water future: The Metropolitan Water Plan 2004*. NSW Department of Infrastructure, Planning and Natural Resources, October 2004, www.dipnr.nsw.gov.au.

Dunmore-Lang J. (1834). *A Historical and Statistical Account of New South Wales*. Dr John Dunmore-Lang, London, 1834.

Restoration of Degraded River Basins: Summary of Workshop Discussion

John Tyson

45 Riverside Close, Warrington, WA1 2JD, UK (john.m.tyson@lineone.net)

In this summary I have tried to draw together the main points of the discussion. I have relied on the recording of the discussion made by Avi Ostfeld to whom my thanks. I hope that readers find it a useful addition to the papers and presentations.

It is, I believe, something of a first for IWA for a discussion to be published in this way and, on behalf of the Watershed & River Basin Management Specialist Group I want to express our thanks to all those who have made it possible.

The discussion was chaired by Alan Vicory; the five authors constituted the panel. It became apparent that, although the problems in the five catchments differed in nature and complexity there was a common view on the development of solutions.

The panel strongly endorsed the principle of the catchment management approach and the development of a catchment based strategy to achieve desired changes and improvements in the water quality and flow regime of a river basin. The manner in which countries may adopt catchment management will differ according to their own circumstances but will typically include both regulatory and institutional means.

In Europe the Water Framework Directive can be considered as providing a regulatory umbrella which requires the development of a River Basin Management Plan for all river basins within the EU, and also establishes the criteria for the involvement of stakeholders. The Mersey Basin Campaign, whilst not a regulatory body, does need to take a comprehensive view of the basin, and engage extensively in consultation and public participation in order to work towards it's objectives.

In Australia, the New South Wales Water Management Act, 2000 with its key objectives of:

- Water management for sustainable use;
- New water access framework to prevent over-allocation;
- Water sharing plans to share available water between users and environment,led to the recent (2003) establishment of 35 Catchment Management Authorities, one for each of the major river basins in New South Wales.

South Africa has established, under the National Water Act 1998, 19 Water Management Areas with the intention that, over time, each of these will become a Catchment Management Agency responsible for the management of water resources in its area.

In Israel, the Alexander river restoration project is, in contrast to the other projects, carried out by a non-institutional, voluntary body. Much stress was placed on the need to establish the trust of the people involved in the catchment and to develop the practice of working together by agreement. It was agreed by all that the need to work together with the catchment

 River Basin Restoration and Management edited by Avi Ostfeld and John M Tyson. ISBN: 184339510X. Published by IWA Publishing, London, UK.

community and to gain, and keep, their trust, is key to the development and implementation of any catchment plan.

An important feature of the catchment management approach is the creation of a long-term vision, incorporated in a 'master plan' or a 'river basin plan'. Common to the creation of such plans is emphasis on the need for the identification of 'stakeholders' (the public, industries, agriculture, local authorities, NGOs) in the catchment to work together with the catchment agency on development and implementation of the plan. The need to combat eutrophication in the Black Sea by control of nutrient inputs, to the Danube, many kilometres upstream provides a good example of the need for, and value of, a river basin plan. Without the catchment based approach it is possible that the solution would not be identified, let alone agreed.

Also emphasised was the importance of a master plan as a living document – not to be put on a shelf and left to gather dust, but a working document that can, and should, be reviewed as circumstances change with time. In South Africa, for example, the Water Act requires the production, every 4 years, of a Water Service Plan, which includes water sources, water demand and catchment management strategy.

Asked about current and future concerns for catchment management and how those concerns might be mitigated, the panel gave a variety of individual responses:-

- the impact of land use changes on river quality and ecology and thus the need to include land use planning within catchment management;
- climate change with potential impact for more severe and longer duration droughts with possible increase in salinity in some cases;
- increasing efficiency in agriculture leading to 'water allocation surplus' and a market in 'allocation transfer';
- increases in commercial navigation as part of reduction in 'greenhouse gases';
- additional power plants utilising water for power generation of cooling purposes to meet increasing energy demands;
- integration of groundwater and surface water flows/quality and the future of wetlands;
- river catchment administration in some countries is often very fragile with no long-term budget and is susceptible to political whim;
- need for social/behavioural changes by the general public requiring more/better awareness and educational programmes.

The final statement of the discussion was a plea for the widespread application of the catchment approach – 'what everyone does should be within a catchment approach!'.

River Basin Management using Machine Learning

Watershed management – a physically based approach versus data driven modeling

A. Ostfeld *

* Department of Civil and Environmental Engineering, Technion – Israel Institute of Technology, Haifa 32000, Israel (Email: *ostfeld@tx.technion.ac.il*)

Abstract This paper describes briefly and compares the two basic approaches to model hydrological phenomena in watersheds: physically based versus data driven modeling. The physically based approach seeks to capture the system physics through sets of equations or storage and routing elements (e.g., linear reservoirs, linear canals) connected together in a specific form. On the contrary, data driven modeling seeks to capture the watershed behavior empirically by employing data driven models (e.g., artificial neural networks). These two approaches represent two extreme attitudes towards modeling. These are discussed and compared.

Keywords Watershed, management, physically based, data driven, modeling, system.

INTRODUCTION

Hydrological phenomena are one of the most complex processes to describe constituting air, soil, and water, interacting together on different spatial and temporary scales. This paper describes briefly and compares the two basic approaches to model hydrological phenomena in watersheds: physically based versus data driven modeling, both within a perspective of watershed system management. Within the perspective of watershed system management the watershed is treated as a system (i.e., an entity which operates on input signals and produces output signals), for the purpose of reconstructing past events, predicting future events, or managing scenarios within the watershed itself.

A physically system approach to determine outputs from a given system requires descriptions of: (1) the system input (s), (2) the system structure, (3) the physical laws which govern the system behavior, (4) boundary conditions, and (5) initial conditions. This can be expressed symbolically as:

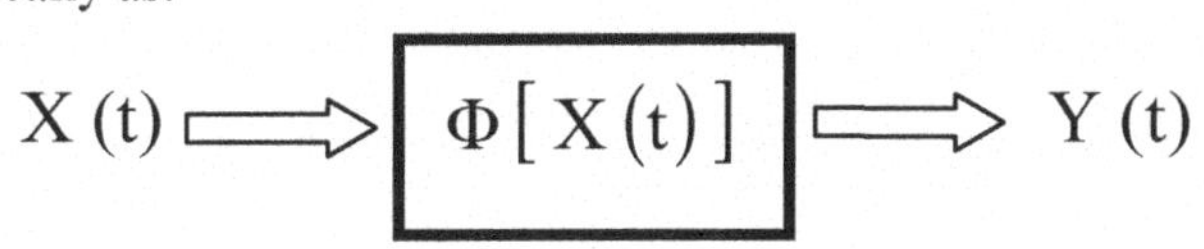

Fig. 1: A schematic system approach

where: X (t) = input (s); $\Phi\left[X(t)\right]$ = the system transformation function of the input (s) X (t) to its output (s) Y (t). The characteristics of Φ define the system nature: linear, non-linear, lumped, distributed, deterministic, or stochastic. Physical modeling concentrates on constructing Φ using the physical understanding of the system behavior.

 River Basin Restoration and Management edited by Avi Ostfeld and John M Tyson. ISBN: 184339510X. Published by IWA Publishing, London, UK.

Contrary to the physically system approach Data Driven Modeling (DDM) (or Machine Learning) seeks to construct Φ empirically by learning the data "behavior" of given sets of X (t) and Y (t). The knowledge learned from the "behavior" of X (t) and Y (t) is built into a DDM engine [i.e., an Artificial Neural Network (ANN), a Model Tree (MT)] for further predictions.

The sections below summarize briefly the two approaches.

PHYSICALLY BASED MODELING

A physical based approach expresses a primarily fundamental motivation to study the physical phenomena underlying physical processes, striving for explanations of how things happen and why they behave as they do.

Klemes (1972) defined three basic phenomena a physically hydrological based model should incorporate:

(1) *Independency.* A hydrological model must be independent of any intended application. In other words, it must be possible to construct a reasonable model of a hydrological process without knowing in advance for what it will eventually be used. In nature a given process behaves as it does regardless of any intentional use.

(2) *Process compatibility.* A hydrological model must incorporate – or at least be compatible with – the known physical structure and mechanism of the whole modeled process. If the model is aimed at and accordingly designed for an explanation or description of only one particular aspect of the process, it is very likely to end up on a dead end.

(3) *Confidence limits.* Observed hydrological process must be within the confidence limits derived for them from the model. The more parameters fall "out of range", the less credibility the model commands.

Numerous papers have been published on using the physically based approach for watershed system management at the last seven decades; as well as ways to classify those models. One of those ways to classify physical models for watershed management is shown in Fig. 2.

<table>
<tr><td></td><td colspan="3">Time variant</td><td colspan="3">Time invariant</td></tr>
<tr><td rowspan="2">Lumped models</td><td>Linear</td><td>Quasi-linear</td><td>Non linear</td><td>Linear</td><td>Quasi-linear</td><td>Non linear</td></tr>
<tr><td colspan="3">Deterministic / Stochastic</td><td colspan="3">Deterministic / Stochastic</td></tr>
<tr><td rowspan="2">Distributed models</td><td>Linear</td><td>Quasi-linear</td><td>Non linear</td><td>Linear</td><td>Quasi-linear</td><td>Non linear</td></tr>
<tr><td colspan="3">Deterministic / Stochastic</td><td colspan="3">Deterministic / Stochastic</td></tr>
</table>

Fig. 2: Model classification for watershed management

Traditionally the watershed was treated as a lumped, time invariant, linear, deterministic system, resulting in the unit hydrograph theory (Sherman, 1932), upon which the Nash (Nash, 1957) linear cascade of reservoirs model, and the parallel cascade of reservoirs of Diskin (1964) were built upon. Other models within this category are attributed to Dooge (1959), Diskin and Boneh (1975), Eagleson et al. (1966), and others. Later models have expanded the lumped linear deterministic approach to a distributed linear cell approach in which the entire watershed is partitioned into a tree like structure built of cells, with each cell been a sub-watershed (Diskin et al., 1984).

The Geographical Information System (GIS) technology stepped in for watershed management a decade ago, providing numerous tools to support hydrologic modeling. These can be broadly classified as data management (e.g. manipulation, preparation, extraction), visualization, and interface tools development. GIS is thus capable of describing hydrological features that are available in a watershed on a spatial scale. Hence, GIS and hydrologic modeling represent two entirely different aspects: description and prediction, respectively. A GIS based hydrological model couples these two in a single one physical framework. Examples for these are SWAT – a river basin scale model developed to quantify the impact of land management practices in large, complex watersheds; BASINS – a multipurpose environmental analysis system for performing watershed and water quality based studies, and AVGWLF – a spatial distributed watershed model based on GWLF (Haith and Shoemaker, 1987) for simulating runoff, sediment, and nutrient (N and P) loadings from a watershed given variable-size source areas (e.g., agricultural, forested, and developed land).

DATA DRIVEN MODELING

The rapid advance in information processing systems and the data collection capabilities in recent decades have directed engineering research towards the development of intelligent systems that can evolve models of natural phenomena automatically. This is the discipline of Data Driven Modeling (DDM) [or Machine Learning (ML)] which is the study of computer algorithms that improve automatically through experience. Applications of DDM range from data-mining programs that discover general rules in large data sets, to information filtering systems that automatically learn users' interests.

The most common techniques of DDM's are: Decision Trees (DT), Model Trees (MT), Artificial Neural Networks (ANN), Bayesian Methods (BM), Fuzzy-Rule (FR) based systems, Support Vector Machines (SVM), and, to some extent, Evolutionary Algorithms [e.g., Genetic Algorithms (GA)]. Since the last decade ML techniques in both hydrology and water resources engineering are gaining an increasing interest.

Numerous papers have been published in the research literature during the last ten years on watershed hydrology and water resources. Some of those are briefly cited below.

Villafana et al. (2004) presented several pre-processing procedures on data for real time flood forecasting using Artificial Neural Networks, including the influence of the watershed wetness, the rainfall characteristics, and a cross correlation routine applied to the rainfall and runoff in order to determine the time steps required to shift one series in relation to the other, for maximizing the mutual correspondence between rainfall and runoff prior to entering the ANN model. Muttil and Liong (2004) applied Genetic Programming (GP) to generate empirically the underlying equations which connect input to output for the Upper Bukit Timah (UBT) watershed in Singapore. Dorado et al. (2003) proposed an application of genetic programming (GP) linked with artificial neural networks (ANN) for rainfall-runoff modeling. The system was designed to serve as a real-time alarm tool against floods in urban basins. Solomatine and Dulal (2003) investigated the accuracy performance of artificial neural networks (ANN) versus model trees (MT) for rainfall-runoff modeling, showing high performances results for both, with a slight advantage given to model trees. Hewett (2003) explored the problem of identifying predictive relationships that can be used to integrate solar and ocean-atmospheric conditions into forecasts of regional water flows comparing neural networks, decision tree learning, and associational rule mining. Abebe et al. (2000) applied a fuzzy ruled based approach to reconstruct missing precipitation events. The working rules were formulated from a set of past observations using an adaptive algorithm. The model was applied to a case study in Northern Italy. Castano et al. (1998) used a supervised classification of images that combined maximum likelihood algorithms and decision-tree criterions for evaluating extracted water volumes needed for irrigation. Lohani and Loganathan (1997) formulated a non-homogeneous Markov chain model to characterize the stochastic behavior of drought. The computed probabilities were used to develop a decision tree for drought management, enumerating all possible sequences of drought occurrences. Trauth and Xanthopoulos (1987) used decision trees to describe the relationships between non-point sources of anthropogenic pollutants in the urban areas and measured concentrations in groundwater.

COMPARISON and CONCLUSIONS

Fig. 3 is a subjective general comparison of the author for the two approaches with respect to a set of eight modeling characteristics. The score of 5 is given to a specific attribute under a specific approach, if it corresponds to "high", "easy", or "cheap"; the score of 1 if it corresponds to "low", "difficult", or "expensive"; and 2, 3, or 4 in between.

Attribute	Physically based modeling	Data driven modeling
1. Insight	5	1
2. Calibration / parameterization	2	5
3. Ability to consider spatial and temporal variations	3	2
4. Flexibility	4	1
5. Robustness	3	3
6. Construction ease	2	5
7. Construction cost	3	1
8. Prediction accuracy	3	5
9. Design prediction capabilities	4	1
Total score	29	24

Legend: 5 = high, easy, cheap; 1 = low, difficult, expensive

Fig. 3: A subjective comparison between a physically based versus data driven modeling approach

It can be seen from Fig. 3 that physically based modeling inherently suppress Data Driven Modeling (DDM) in terms of insight and flexibility as a physically based model relies on capturing the physics of the system while a data driven model – on available data. Also, a DDM is unable to predict "if then" design scenarios reflecting modifications in the watershed itself; a DDM is capable of exploring only modifications made to the data it was trained for.

On the other hand it is extremely fast and easy to construct a DDM once the data and a DDM engine (e.g., an Artificial Neural Network, a Model Tree) are available. The latter is especially true if the data is indeed available. If this is not the case then constructing a DDM can be extremely expensive, and in many cases completely impossible (e.g., if there is a need to collect annual data).

The other attributes in Fig. 3 depend much more on the specific case of interest, thus it is much harder to draw a definite statement.

Looking at the total score of both approaches leads to the conclusion that there is no definite answer as of what approach is superior. Things are case dependent and thus each approach might suppress the other, depending on the specific problem and the data in hand. This general conclusion might call for a hybrid approach in which part of the model is physically based (e.g., series of linear reservoirs), capturing parts of the physics; while the other are data driven. Little has been reported in the research literature with respect to that.

REFERENCES

Abebe A. J., Solomatine D. P., and Venneker R. G. W. (2000). Application of adaptive fuzzy rule-based models for reconstruction of missing precipitation events. *Hydrological Science Journal*, Vol. 45, No. 3, pp. 425 – 436.

AVGWLF. *http://www.avgwlf.psu.edu/*

BASINS. *http://www.epa.gov/waterscience/basins/basinsv3.htm*

Castano M., Mejuto M. F., and Calera A. (1998). Application of remote sensing techniques to the evaluation of water consume. *The International Society for Optical Engineering,* Vol. 3499, pp. 58-68.

Diskin M. H. (1964). A basic study of the linearity of the rainfall – runoff process in watersheds. *PhD Thesis*, University of Illinois, Urbana, Champaign, USA

Diskin M. H. and Boneh A. (1975). Determination of an optimal IUH for linear time invariant systems from multi – storm records. *Journal of Hydrology*, No. 24, pp. 57 – 76.

Diskin M. H., Wyseure G., and Feyen J. (1984). Application of a Cell Model to the Bellebeek Watershed. *Nordic Hydrology*, No. 15, pp. 25 – 38.

Dooge J. C. I. (1959). A general theory of the unit hydrograph. *Journal of Geophysics Research*, Vol. 64, No. 2, pp. 241 – 256.

Dorado J., Puertas J., Santos A., Rivero D., Pazos A., Rabunal J. R. (2003). Prediction and modeling of the rainfall-runoff transformation of a typical urban basin using ANN and GP. *Applied Artificial Intelligence*, Vol. 17, No. 4, pp. 329-343.

Eagleson P. S., Mejia R., and March F. (1966). Computation of optimum realizable unit hydrographs. *Water Resources Research*, Vol. 2, No. 4, pp. 755 – 764.

Haith, D. A. and Shoemaker L. L. (1987). Generalized Watershed Loading Functions for Stream Flow Nutrients. *Water Resources Bulletin*, Vol. 23 No. 3, pp. 471-478.

Hewett R. (2003). Data Mining for Generating Predictive Models of Local Hydrology. *Applied Intelligence*, Vol. 19, No. 3, pp. 157-170.

Klemes V. (1972). Comment on Mathematical models of hydrologic systems by J. C. I. Dooge. *Proceedings of the International Symposium on Modeling Techniques in Water Resources Systems*, Ottawa, Canada, Vol. 3, pp. 701-705.

Lohani V. K. and Loganathan G. V. (1997). Decision tree approach for drought management. *Proceedings of the Annual Water Resources Planning and Management Conference*, Aesthetics in the Constructed Environment, pp. 100-105.

Muttil N. and Liong S. Y. (2004). Physically interpretable rainfall-runoff models using genetic programming. *The 6th International Conference on Hydroinformatics* - Liong, Phoon and Babovic (Eds.), published on CD.

Nash J. E. (1957). The form of the instantaneous unit hydrograph. *IASH Publication* No. 45, Vol. 3, pp. 114 – 121.

Sherman L. K. (1932). Stream flow from rainfall by the unit graph method. *Engineering News-Record*, Vol. 108, pp. 501 – 505.

Solomatine D. P. and Dulal K. N. (2003). Model trees as an alternative to neural networks in rainfall-runoff modeling. *Hydrological Science Journal*, Vol. 48, No. 3, pp. 399 – 411.

SWAT. *http://www.brc.tamus.edu/swat/index.html*

Trauth R. and Xanthopoulos C. (1987). Non-point pollution of groundwater in urban areas. *Water Research*, Vol. 31, No. 11, pp. 2711-2718

Villafana G. C., Cluckie I. D., and Han D. (2004). An approach on evaluating the pre-processing of rainfall and runoff data in an artificial neural network model for real –time forecasting. *The 6th International Conference on Hydroinformatics* - Liong, Phoon and Babovic (Eds.), published on CD.

A Fuzzy Method of Assessing Sustainability in Stormwater Planning

L Cobiac*, K Jansons**

*Cooperative Research Centre for Coastal Zone, Estuary and Waterway Management and Department of Mathematics, The University of Queensland, St Lucia, 4072, Australia
**Division of Environment, School of Engineering, The University of Queensland, St Lucia, 4072, Australia

Abstract New modelling tools are required if we are to achieve sustainable management of urban stormwater resources. But sustainability is a complex concept that encompasses a variety of social, economic and ecological issues of which we frequently have insufficient knowledge and data to model using traditional stormwater modelling methods. Fuzzy systems are a useful method of modelling complex systems, and in this paper we present a tool that has been developed for assessing sustainability of stormwater management strategies using a fuzzy systems approach. The research illustrates the benefits of taking a fuzzy systems approach to modelling problems where imprecision is a challenge.

Keywords Sustainability assessment; Urban stormwater; Fuzzy logic

Introduction

Sustainable management of stormwater resources is vital to the future of our urban communities and ecological systems. To achieve a more efficient and equitable use of our water resources, we need ways of assessing the sustainability of our management actions so that we can better define and measure progress towards sustainability goals.

Since the nineteenth century, urban planners have relied on mathematical models to help determine the best management actions. These models have been developed primarily using stochastic and deterministic modelling techniques. The accuracy and precision has improved over time as knowledge and data on climate, river hydraulics and water quality has advanced. Today, there are many mathematical models available to assist stormwater managers in their planning, and they are widely used for assessing potential impacts of management actions on the flow and water quality characteristics of waterways.

These more conventional methods of constructing mathematical models, however, are not easily extended to address the broader concept of sustainability. Bell and Morse (1999) have aptly described sustainability as "a qualitative property of a system". It is a vague and complex concept, encompassing social, economic and ecological issues, many of which are challenging to understand and challenging to measure. Sustainability is not amenable to conventional mathematical modelling methods, and thus there has been growing interest in the use of alternative modelling methods for its assessment.

Fuzzy systems are a precise method of working with imprecise information, and are a potential alternative. The concept first appeared in the 1920s (Durkin, 1994), but it was the work of Zadeh (1965) that transformed it into a practical system of mathematical logic, and it was another 5 years before the method was applied to decision-making problems by Bellman

 River Basin Restoration and Management edited by Avi Ostfeld and John M Tyson. ISBN: 184339510X. Published by IWA Publishing, London, UK.

and Zadeh (1970). Since this time, it has been widely used in places like Japan to design smarter consumer products, such as cars and washing machines. Only recently, however, have fuzzy methods been applied to sustainability problems.

Fuzzy systems are based on fuzzy set theory, which is an extension of classical set theory. In fuzzy set theory objects may take on *partial* membership of a set rather than either belonging or not belonging as in the classical case. To represent this mathematically, we let $X = \{x\}$ be a collection of objects, where a generic element of X is denoted by x, and define a fuzzy set A in X.

$$A = \{(x, \mu_A(x))\} \qquad \text{for } x \in X$$

This is a set of ordered pairs where $\mu_A(x)$ is the membership grade of x in A. In conventional set theory, $\mu_A(x)$ could take only the value 0 (i.e. x does not belong to A) or 1 (i.e. x does belong to A), but in fuzzy set theory, $\mu_A(x)$ can be any real number in the interval [0, 1].

When this basic theory is applied to logic problems, the classical Boolean logic of true and false can be extended to multi-valued logic involving degrees of truth. This fuzzy logic gives computers the ability to mimic the human capacity to reason imprecisely and make judgements under uncertainty. Thus it is a potentially powerful tool for sustainability assessment.

Researchers have been exploring the potential of fuzzy systems using a variety of methods and in different sustainability applications. Ducey and Larson (1999), for example, have used fuzzy set theory to evaluate and compare forest management decisions. Cornelissen et al. (2001) investigated sustainable agricultural development using aggregation operations and approximate reasoning. While Phillis and Andriantiatsaholiniaina (2001) approached sustainability from a global perspective, using fuzzy logic to measure the sustainability of national economies. There are many possible methods of applying fuzzy systems to sustainability problems.

Imprecise reasoning is the primary advantage of fuzzy systems in sustainability applications, but they also have an advantage over more conventional modelling methods due to their linguistic capacity. The use of linguistic terms from our natural language, such as *very big* or *somewhat good*, make it possible to use information in a sustainability assessment that is difficult to measure in numerical terms. Thus there are potentially multiple advantage of a fuzzy systems approach.

In our research we have been investigating the use of a fuzzy systems approach to assessing sustainability in stormwater planning. We have developed a method for assessing sustainability of stormwater management strategies using fuzzy systems, and have applied it to an urban catchment on the east coast of Australia (Figure 1.1). In this paper, we present the results of this work, and discuss the research implications for the development of fuzzy systems tools to achieve sustainable management of water systems.

A framework for sustainable stormwater management

The assessment of sustainability is based on an indicator assessment process. Sustainability objectives are first defined for the catchment shown in Figure 1.1, assessment criteria are then chosen, and indicators selected to provide information for the assessment.

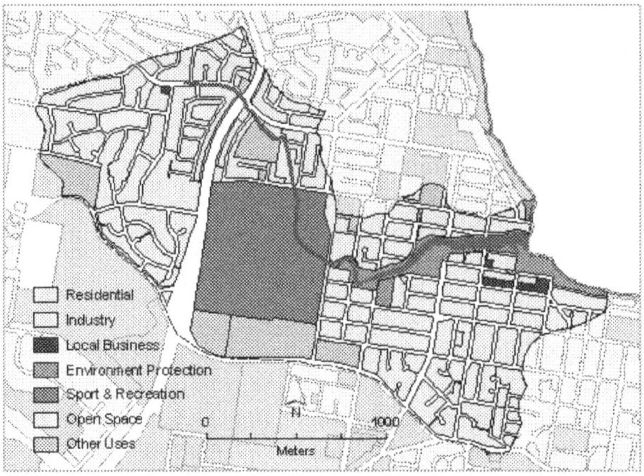

Figure 1.1 The urban catchment

For stormwater management, there are two key objectives to achieving healthy and self-sustaining urban communities. The first objective is to achieve a balance between human use of freshwater resources and disposal of polluted stormwater wastes, and the environment's capacity to provide those resources and assimilate the wastes. This is a challenging task, given that this 'carrying capacity' is not known, nor is it likely to be static. Thus the second objective is to maximise the potential for achieving progress in the sustainable direction.

A number of criteria have been identified for the catchment shown in Figure 1.1, which can be used to assess the extent to which stormwater management strategies meet the objectives. The key criteria include a healthy ecosystem, meeting people's basic needs for health and safety, the availability of sufficient means of implementing the management plan (including finance, expertise, labour and resources), and providing individuals with the means and the motivation to carry out on-going management responsibilities. These key criteria are further broken down into sub-criteria, forming the hierarchy shown in Figure 2.1. Indicators provide the input information to the assessment and form the lowest level of the hierarchy.

Linguistics and membership functions

Linguistic terms are defined to describe each of the variables in the framework shown in Figure 2.1. The linguistics used to define variables at Levels 1 to 3 are *very bad*, *bad*, *satisfactory*, *good* and *very good*, and the linguistics used to define variables at Levels 4 and 5 are *low*, *medium* and *high*. Each of the linguistics is a fuzzy set. Their shape is described by the membership functions shown in Figure 3.1.

Level 5 variables receive direct input of indicator information, and linguistic hedges are used to provide greater descriptive power for input to these variables. Hedges are pre-modifiers, such as *very* and *somewhat*, that can be used to modify the linguistic adjectives. Mathematically, they modify the shape of the linguistic fuzzy set. Adding the hedge *very* to the linguistic term *low*, for example, concentrates the fuzzy set describing the term *low*, while adding the hedge *somewhat* has a dilating effect.

$$\mu_{very\,low}(x) = (\mu_{low}(x))^2$$

$$\mu_{somewhat\,low}(x) = (\mu_{low}(x))^{0.5}$$

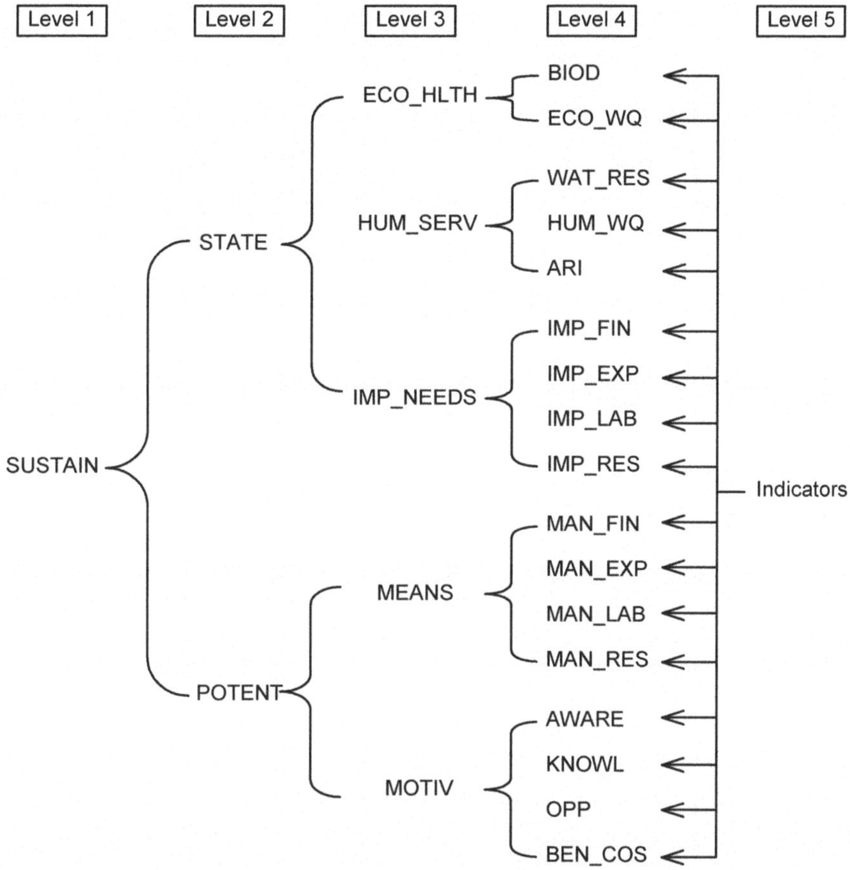

SUSTAIN – Sustainability; STATE – State of system; POTENT – Potential for improvement; ECO_HLTH – Ecosystem health; HUM_SERV – Human services; IMP_NEEDS – Meeting needs for implementation of scenario; MEANS – Providing means for on-going operation & maintenance; MOTIV – Motivation for sustainable behaviour; BIOD – Biodiversity; ECO_WQ – Water quality for protection of aquatic ecosystem; WAT_RES – Re-use & recycling of water; HUM_WQ – Water quality for primary contact; ARI – Flooding risk; IMP_FIN – Finance available for implementation of scenario; IMP_EXP – Expertise required for implementation of scenario; IMP_LAB – Labour required for implementation of scenario; IMP_RES – Resources required for implementation of scenario; MAN_FIN – Finance available for on-going management; MAN_EXP – Expertise required for on-going management; MAN_LAB – Labour required for on-going management; MAN_RES – Resources required for on-going management; AWARE – Awareness of problem; KNOWL – Knowledge & skills to act; OPP – Opportunity to participate; BEN_COS – Benefits versus costs of participating

Figure 2.1 Sustainability assessment framework

Input information may be either crisp or fuzzy. Crisp inputs (e.g. x_i) are single numbers, and can be used where specific indicator data are available. For variables that are vague or imprecise, fuzzy inputs provide a more honest representation. Fuzzy inputs are in fact fuzzy set themselves (e.g. $\mu_{A_i}(x)$), and input values are determined by the intersection of sets.

$$\begin{aligned}\mu_{A\cap A_i}(x) &= \mu_A(x) \wedge \mu_{A_i}(x) \qquad \text{for } x \in X \\ &= \min\{\mu_A(x), \mu_{A_i}(x)\}\end{aligned}$$

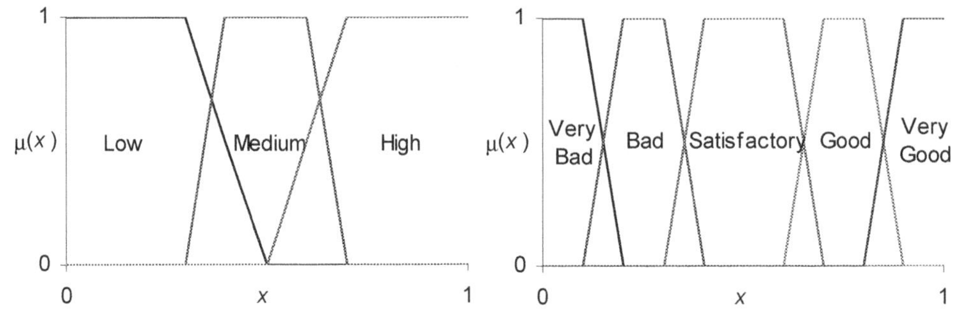

Figure 3.1 Examples of membership functions for (a) Level 4 and 5 variables; and (b) Level 1, 2 and 3 variables

Fuzzy rule-based system

The sustainability framework (Figure 2.1) provides the structure for the rule relationships and forms the procedural basis for the fuzzy calculations. Each level of the framework is linked to its adjacent level by a set of rules. These rule bases represent our current state of knowledge about sustainability.

Each fuzzy rule base is a collection of rules based on the form IF [antecedent] THEN [consequent], where each fuzzy rule relates one proposition to another.

IF x is A THEN y is B *or* IF A THEN B

The fuzzy relationship, R, for the rule IF A THEN B, or R = A → B, relates the antecedent and consequent parts of the rule. This is defined by the Mamdani implication, which produces a two-dimensional membership function for the rule relation.

$$\mu_R(x, y) = \min\{\mu_A(x), \mu_B(y)\}$$

Since the rule-based system contains more than one rule, the overall outcome or consequent must be determined by aggregation of the rules. This calculation depends on whether the rules are conjuctive (connected by AND) or disjunctive (connected by OR). For rules that are conjunctive, all rules must be satisfied, and the overall outcome is found by the fuzzy intersection of the consequents of the individual rules.

$$\mu_A(y) = \min\{\mu_{A_1}(y), \mu_{A_2}(y), \ldots, \mu_{A_n}(y)\} \qquad \text{for } y \in Y$$

For rules that are disjunctive, at least one of the rules must be satisfied, and the overall outcome is found by the fuzzy union of the consequents of the individual rules.

$$\mu_A(y) = \max\{\mu_{A_1}(y), \mu_{A_2}(y), \ldots, \mu_{A_n}(y)\} \qquad \text{for } y \in Y$$

Thus, illustrating the Mamdani implication for n disjunctive rules with 2 antecedents, the aggregated output is given by:

$$\mu_{B^r}(y) = \max_r\{\min\{\mu_{A_1^r}(x_{i1}), \mu_{A_2^r}(x_{i2})\}\} \qquad \text{for } r = 1, 2, \ldots, n$$

The centroid method of defuzzification is applied to the induced fuzzy set $\mu_{B^r}(y)$ to obtain a single crisp value, y^*.

$$y^* = \frac{\int \mu_{B_r}(y).y\,dy}{\int \mu_{B_r}(y)\,dy}$$

This procedure is carried out for each variable shown in the indicator framework (Figure 2.1), starting with the lowest level variables at Level 5 and ending with the highest level variable, the overall sustainability index at Level 1. It is illustrated graphically for a single variable in Figure 4.1.

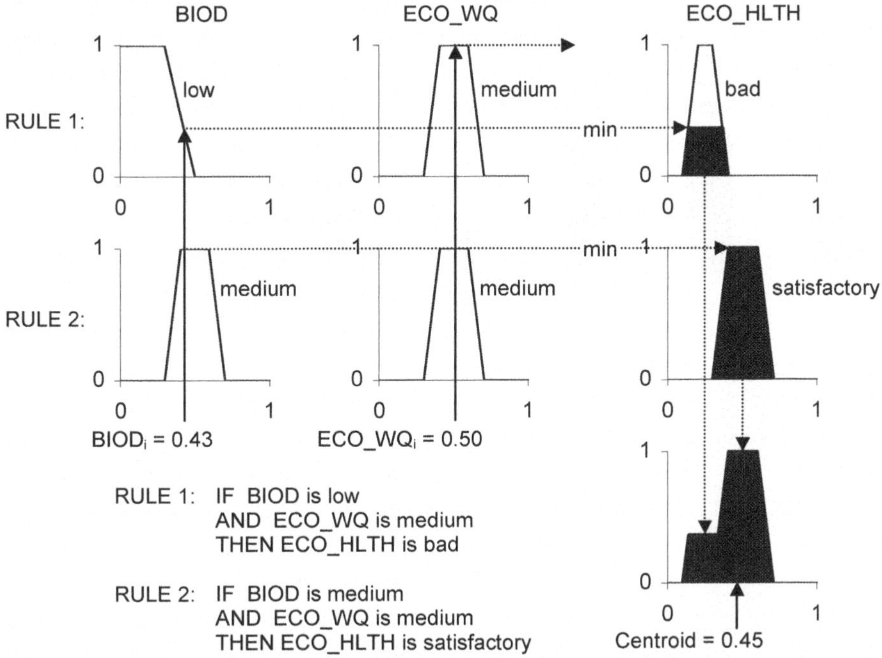

Figure 4.1 Graphical illustration of the fuzzy method for ECO_HLTH

Sustainability assessment using the fuzzy rule-based system

The fuzzy rule-based system was implemented for the catchment shown in Figure 1.1 to assess the sustainability of stormwater management strategies currently in place. The fuzzy inputs were based on indicator information derived from a number of sources including monitoring data (e.g. water quality) and expert judgement. An overall sustainability index of 25 % was determined.

Figure 5.1 illustrates the outputs for the Level 3 variables. Comparison of these variables is helpful in identifying strengths and weaknesses of stormwater management in the catchment. In this case, for example, current methods of stormwater management appear to be weaker in providing people with the means and motivation to keep progressing towards a more sustainable state. This is generally the result of a reliance on end-of-pipe strategies, such as gross pollutant traps and sedimentation basins, and can be remedied by greater investment in preventative strategies, such as education to encourage more sustainable behaviour.

The shape of the membership functions was found to be critical to the sustainability results. Modification of the membership function for Level 1, 2 and 3 variables, as shown in Figure 5.2, for example, increased the sustainability index from 25 % to 46 %.

Membership functions are defined intuitively for the sustainability variables. Although defined according to the best available knowledge, they will inevitably be subjective and specific to the catchment.

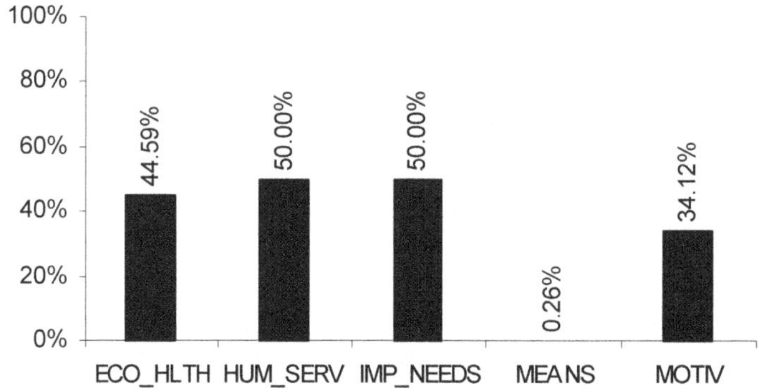

Figure 5.1 Fuzzy outputs for level 3 variables

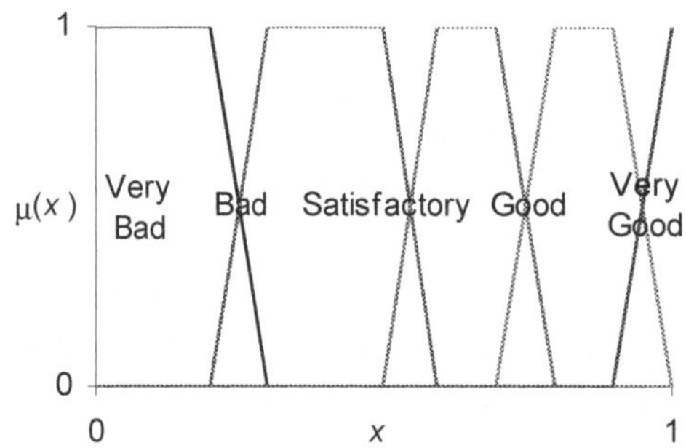

Figure 5.2 Modified membership function for Level 1, 2 and 3 variables

The sensitivity does not detract from the use of the method for evaluating stormwater management strategies for the catchment to which it is applied, or its ultimate intended use as a means of assessing hypothetical management scenarios to identify the most sustainable practices. But it does highlight the importance of ensuring the method is not mis-used for making catchment comparisons, or as a policy tool for 'rating' the sustainability of different catchments.

The fuzzy rule-based system is not a precise method. If sustainability was a simple well-defined concept it would be easy to adopt a more traditional equation-based mathematical modelling method. Even in the absence of understanding, if variables were quantifiable and data available, artificial neural network type methods could be a good alternative. But sustainability is a complex concept involving variables that are imprecise and difficult to quantify, and is thus suited to a fuzzy systems approach.

As a method, fuzzy systems are not easy to validate in the traditional sense. The sustainability index and its composite variables are not characteristics like water quality parameters, for example, which can be measured and compared. The fuzzy system is a means of approximately reasoning the outputs based on the information with which it is supplied. This information not only includes the indicator inputs, but also the expertise captured in the rules and the shapes of the membership functions. Thus it cannot be validated the same way a more traditional method could be.

The key advantage that was found in taking the fuzzy systems approach was the ability to incorporate *all* stormwater management decision criteria. Existing models focus on thos criteria that are precise and quantifiable, and can lead to an undervaluation of other criteria, particularly those relating to the social changes that are important for achieving greater sustainability in the long term. The fuzzy systems approach encourages a more balanced assessment of the sustainability of stormwater management plans.

A second advantage of the fuzzy systems approach over other modelling methods, is the potential ease of updating the system and adapting it to new applications. Knowledge of the system is held in the rules and the shapes of the membership functions, which can be easily updated by the user. There is no need for the developer to re-write code for the system to be based on the latest available knowledge. Similarly, the user could apply it to other water systems or systems at different scales (e.g. household scale) just by altering the rules and membership functions. Thus the fuzzy systems approach provides flexibility for the end-users.

Conclusions

The fuzzy system that has been presented in this paper provides a systematic approach to incorporating sustainability objectives into the stormwater planning process for urban planners. The method can be used for evaluating hypothetical stormwater management scenarios to identify strategies that will best meet sustainability objectives for a catchment.

The fuzzy method is ideal for modelling the uncertainty associated with complex systems where the behaviour of the system may be ambiguous, and variables may be imprecise or difficult to quantify. And when combined with a rule-based approach, it becomes a flexible tool that the user can apply to sustainability assessment of different systems and can update with ease to ensure planning is based on the latest knowledge available.

The potential imprecision of the technique would not make it appropriate for every application. But, as illustrated here, for a complex concept such as sustainability it is a valuable addition to the suite of traditional modelling methods which we have available for our decision-making.

References

Bell, S. and Morse, S. (1999) Sustainability Indicators: Measuring the Immeasurable?, Earthscan Publications, London.

Bellman, R. and Zadeh, L. (1970) Decision-Making in a Fuzzy Environment. *Management Science*, **17,** 141-164.

Cornelissen, A.M.G., van den Berg, J., Koops, W.J., Grossman, M. and Udo, H.M.J. (2001) Assessment of the contribution of sustainability indicators to sustainable development: a novel approach using fuzzy set theory. *Agriculture, Ecosystems & Environment*, **86,** 173-185.

Ducey, M.J. and Larson, B.C. (1999) A fuzzy set approach to the problem of sustainability. *Forest Ecology and Management*, **115,** 29-40.

Durkin, J. (1994) Expert Systems: Design and Development. Prentice Hall, Englewood Cliffs, New Jersey.

Phillis, Y.A. and Andriantiatsaholiniaina, L.A. (2001) Sustainability: an ill-defined concept and its assessment using fuzzy logic. *Ecological Economics*, **37,** 435-456.

Zadeh, L. (1965) Fuzzy sets. *Information and Control*, **8,** 338-353.

A data driven model for flow and contaminants runoff predictions in watersheds

A. Ostfeld [*] **and A. Pries** [**]

* Department of Civil and Environmental Engineering, Technion – Israel Institute of Technology, Haifa 32000, Israel (Email: *ostfeld@tx.technion.ac.il*)
** Department of Civil and Environmental Engineering, Technion – Israel Institute of Technology, Haifa 32000, Israel (Email: *preisa@techunix.technion.ac.il*)

Abstract Presented herein is a methodology and application of a hybrid genetic algorithm (GA) - decision trees (DT) - data driven model (DDM) for flow and contaminants runoff predictions in watersheds. The model has two sections: quantity and quality. The quantity section is responsible for finding the decision tree that maximizes flow predictions matching at the watershed outlet; while the quality model uses the quantity optimal outcomes to find the corresponding optimal tree that maximizes contaminants predictions. The model was applied to a sub-watershed of the Lake Kinneret (the Sea of Galilee) basin in northern Israel.

Keywords Decision trees, genetic algorithm, model, rainfall, runoff, watershed

INTRODUCTION

The rapid advance in information processing systems along with the increasing data availability, have directed engineering research towards the development of intelligent systems that can evolve models of natural phenomena automatically. This is the discipline of data driven modeling (DDM) which is the study of computer algorithms that improve automatically through experience.

Applications of data driven modeling (DDM) range from data-mining programs that discover general rules in large data sets, to information filtering systems that automatically learn users' interests.

The most common techniques of data driven modeling are: decision trees, artificial neural networks, Bayesian methods, fuzzy-rule based systems, support vector machines, and, to some extent, evolutionary algorithms. Since the last decade data driven modeling techniques in both hydrology and water resources engineering are gaining an increasing interest. Fig. 1 is a schematic representation of a data driven model framework.

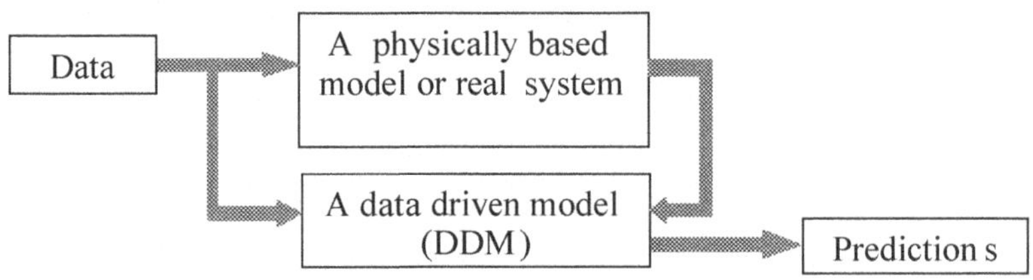

Fig. 1: A schematic data driven model framework

This paper presents the methodology and application of a hybrid decision trees (DT) - genetic algorithm (GA) data driven model (DDM) for flow and contaminant runoff predictions in watersheds.

LITERATURE REVIEW

Numerous papers have been published in the last decade on data driven modeling in hydrology and water resources engineering. Of these few representatives are described below.

Trauth and Xanthopoulos (1987) used decision trees to describe the relationships between non-point sources of anthropogenic pollutants in the urban areas and measured concentrations in groundwater. Lohani and Loganathan (1997) formulated a non-homogeneous Markov chain model to characterize the stochastic behavior of drought. The computed probabilities were used to develop a decision tree for drought management, enumerating all possible sequences of drought occurrences. Castano et al. (1998) used a supervised classification of images that combined maximum likelihood algorithms and decision-tree criterions for evaluating extracted water volumes needed for irrigation. Abebe et al. (2000) applied a fuzzy ruled based approach to reconstruct missing precipitation events. The working rules were formulated from a set of past observations using an adaptive algorithm. The model was applied to a case study in Northern Italy. Hewett (2003) explored the problem of identifying predictive relationships that can be used to integrate solar and ocean-atmospheric conditions into forecasts of regional water flows comparing neural networks, decision tree learning, and associational rule mining. Solomatine and Dulal (2003) investigated the accuracy performance of artificial neural networks (ANN) versus model trees (MT) for rainfall-runoff modeling, showing high performances results for both, with a slight advantage given to model trees. Dorado et al. (2003) proposed an application of genetic programming (GP) linked with artificial neural networks (ANN) for rainfall-runoff modeling. The system was designed to serve as a real-time alarm tool against floods in urban basins.

MODEL DEVELOPMENT

The developed model is a hybrid decision tree (DT) – genetic algorithm (GA) scheme. Decision trees, genetic algorithms, and the resulted hybrid model are next presented.

Decision trees (DT)

Decision tree learning is one of the most widely used and practical method for inductive inference, successfully applied to a broad range of engineering problems.

Prior to implementing a decision tree algorithm, data preparation which include data examination and data splitting into a training portion – required to train and construct a decision tree model, and testing (cross-validation) – needed to validate the model's predictive performances after training, are necessary.

Decision trees classify instances by sorting them down a tree structure from the root to some leaf node. The leaf node provides the instance classification. Each node in the tree represents an attribute of the instance (for example: cold, mild, and hot are discrete instances of the attribute temperature, while 20 ^{0}C, 30 ^{0}C are continuous such instances). Each branch, descending from a tree node, corresponds to one possible value of a specific attribute. An instance is classified by starting at the root node of the tree, testing the attribute specified by that node, and then moving down a branch that corresponds to a selected value of that attribute. Most algorithms that have been developed for decision trees are variations of the basic ID3 algorithm (Quinlan, 1986) summarized in Fig. 2. The ID3 algorithm starts with the question "Which attribute should be tested at the root of the tree?" or - how to identify the best decision attribute. It then continues by selecting the attribute which is the best classifier at a node by maximizing the information gain. The information gain property measures how well a given attribute separates the training examples according to their target classification. The ID3 uses this property to move down the tree.

- A = the 'best' decision element (i.e., the decision attribute that maximizes the information gain).
- Assign A as the decision attribute for the root.
- For each value of A add a new tree branch beneath the root.
- Sort training examples to leaf nodes.
- If the training examples were entirely classified, STOP, else - iterate over new leaf nodes.

Fig. 2: The ID3 algorithm (Quinlan, 1986)

The information gain is defined using Entropy that characterizes the impurity of an arbitrary collection of examples. The entropy for members in S divided into c classes is:

$$E(S) = \sum_{i=1}^{C} -P_i \log_2 P_i \qquad (1)$$

where: P_i = the proportion of examples from the same class. For example if there are 14 instances devided into 2 classes where 9 are positive and 5 are negative, then the entropy is:

$$E([9+, 5-]) = -(9/14)\log_2(9/14) - (5/14)\log_2(5/14) = 0.94 \qquad (2)$$

The main objective is to reduce the total entropy in the system; this is achieved by splitting the set S into subsets with decreasing entropy. Lower entropy values correspond to higher information gain.

If the instances are divided on the basis of attribute A, with the values (A) been the set of all possible values for attribute A, and S_v been the subset of S that has the value v for attribute A, then the information gain (i.e., the expected reduction in entropy caused by knowing the value of attribute A) is:

$$\mathrm{Gain}(S, A) = E(S) - \sum_{v \in \mathrm{values}(A)} \frac{|S_v|}{|S|} E(S_v) \qquad (3)$$

When a given problem is complicated a vast tree will be constructed, resulting the necessity of a pruning procedure to avoid over fitting. The process of pruning is to cut the tree and evaluate the resulted classification accuracy. If the accuracy is kept or improved, then the pruning continues, otherwise the pruning is stopped and the decision tree is converted into an equivalent set of if-then rules by creating one rule for each path, starting from the root node and ending at the leaf node. The most common algorithm for pruning decision trees is C4.5 (Quinlen, 1993), also used herein.

Genetic algorithms (GA)

Genetic algorithms are adaptive search techniques introduced by (Holland, 1975), and further implemented by (Goldberg, 1989). A genetic algorithm is a domain heuristic independent global search technique that imitates the mechanics of natural selection and natural genetics of Darwin's evolution principle. The fundamental idea is to simulate the natural evolution mechanisms of chromosomes (represented by string structures) involving: selection, crossover, and mutation.

A typical GA form involves three main stages: (1) *initial population generation*: the GA generates a bundle of strings, with each being a coded representation of the decision variables; (2) *computation of the strings fitness:* the GA evaluates each string fitness, giving a "fitness-penalty" to infeasible strings; and (3) *generation of a new population*: the GA generates the next population by performing: selection, crossover, and mutation, where: selection involves the process of choosing chromosomes from the current population for reproduction according to their fitness values; crossover involves partial exchange of information between pairs of strings; and mutation – a random change in one of the strings locations. Strings may have binary, integer, or real values.

A typical GA may take the following form:

1. Initialization

1.1 Set the generation counter $t := 0$;
1.2 Generate an initial population $G(0)$;
1.3 Evaluate $G(0)$.

2. Main Scheme

Repeat
2.1 Set: $t = t + 1$;
2.2 Generate $G(t)$ using $G(t-1)$;
2.3 Evaluate $G(t)$;

Until stopping conditions are met.

During the last decade GAs became one of the more robust optimization techniques used to solve engineering problems.

The Hybrid (DT-GA) Model

The hybrid (DT-GA) model is shown in Fig. 3. The model is comprised of two sections: quantity and quality. The objective of the quantity section is to tune the values of a vector of coefficients α that multiply the average rainfall time series intensity I(t) (the input) imposed on a given watershed so as to calibrate its outlet flows Q(t); the quality section then uses these optimal flows Q(t) and the effective optimal rainfall intensities $I_e^*(t)$ to adjust the values of a vector of coefficients β so as to calibrate the watershed outlet concentrations C(t).

The quantity section (top of Fig. 3) comprises the following: t = time; DT = lag in time (e.g., the concentration time of the watershed); α - vector of coefficients (the decision variables); I(t) = time series of the average rainfall intensity imposed on the watershed (e.g., using the average rainfall Theisen method); ET(t) = Evapotranspiration time series; $I_e(t)$ = time series of the effective rainfall; Q(t) = flow time series at the watershed outlet; fitness = the fitness of the decision tree (DT) outcome analysis, estimated through a least square type equation; and GA = genetic algorithm.

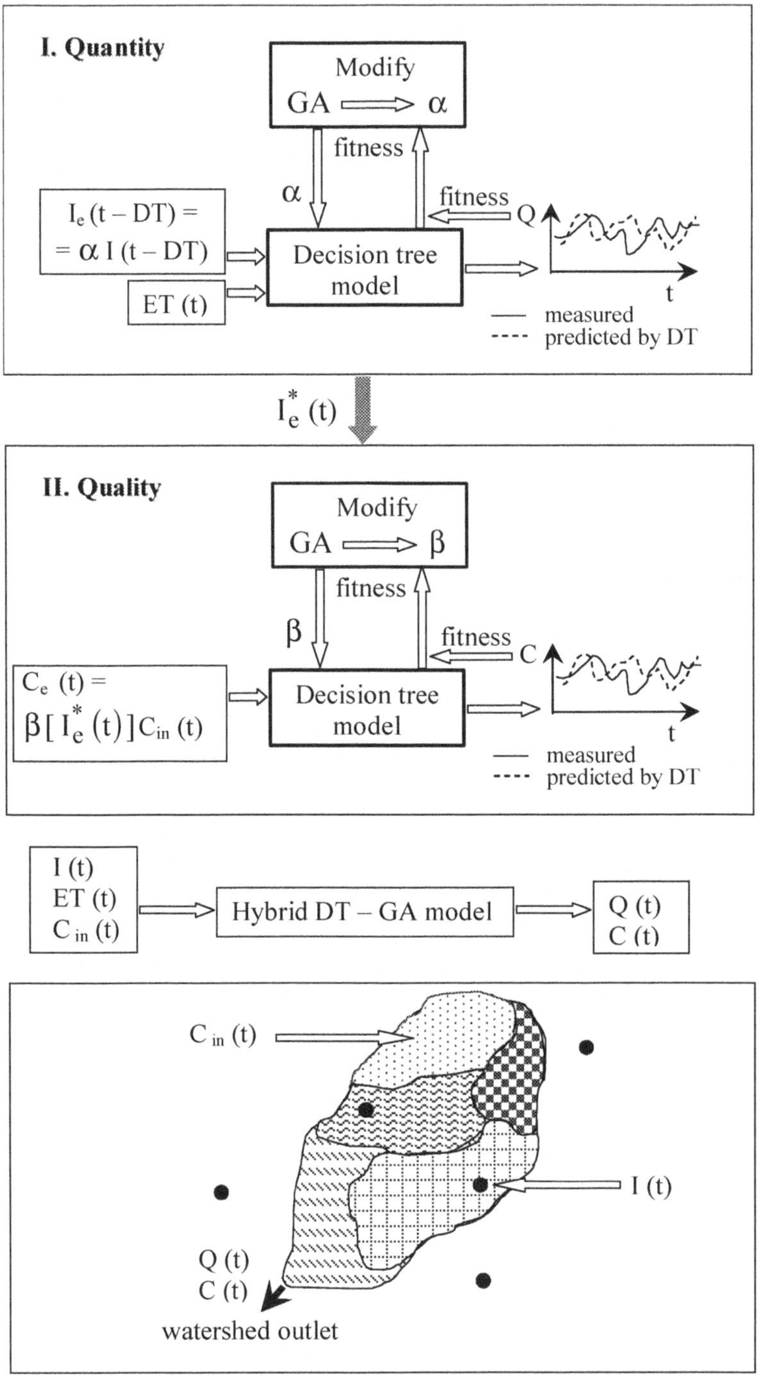

Fig. 3: The hybrid (DT-GA) model

The quality section (mid of Fig. 3) includes the following: $I_e^*(t)$ = the optimal effective rainfall intensity time series (i.e., the outcome of the quantity model); $C_{in}(t)$ = the input concentration time series imposed on the watershed; β - vector of coefficients (the decision variables); $C_e(t)$ = an "effective" resulted concentration time series; C(t) = concentration time series at the watershed outlet; fitness = the fitness of the decision tree (DT) outcome analysis, estimated through a least square type equation; and GA = genetic algorithm.

To reduce the computational complexity and to increase the model robustness, the dimension of the α and β coefficient vectors are set to be much less than the dimension of t. This is accomplished by dividing the time series of the rainfall intensity I(t) to a set of category domains of no more than 6 (i.e., the rainfall intensity is divided into six categories, with α and β values assigned to each). Further to that the number of rules the DT is allowed to establish is constrained to only few (a maximum of 8), so as to receive a robust set of rules (i.e., maximize rules simplicity; minimize over fitting).

APPLICATION

The hybrid (DT-GA) model was applied to Meshushim watershed, a sub-watershed of Lake Kinneret (the Sea of Galilee) basin (Fig. 4). The Lake Kinneret watershed is about 2,730 km^2 [2,070 in Israel, the rest in Lebanon], inhabiting about 200,000 people, organized in 25 municipalities, and three cities (the Israeli part). The Meshushim watershed has an area of about 140 km^2. The soil consists of an upper layer of rocks, clay and sand. The land is mainly used for agriculture. Urban and industrial zones are located at the center of the area around the city of Katzerin. Rainfall in the basin is measured by Mekorot Co. Kinneret Watershed Unit, and by the Israeli Meteorological service on a continuously/daily basis, while monitoring of flow and water quality parameters (e.g., Nitrate, Ammonium, Chloride, Sodium, etc.) on a daily/weekly basis (Markel and Shamir, 2002). The database of contaminants sources (point and non-point) used in this work is that constructed for the Israeli Water Commission by DHV MED (2001). Examples of the model outcomes are shown in Figs. 5 – 7.

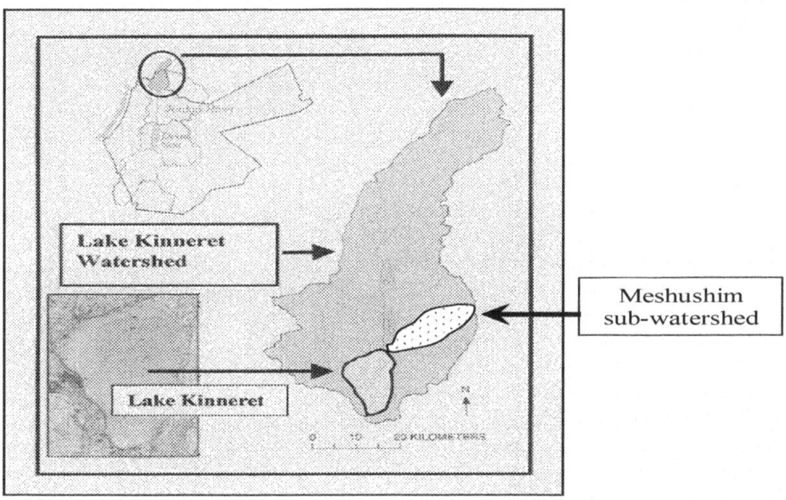

Fig. 4: General layout of Lake Kinneret and Meshushim sub-watershed

Fig. 5 is a typical graph of the convergence performance of the algorithm, both for the quantity and quality sections of the model. The fitness is composed of a maximum value of two, which is the summation of a maximum value of one for the training set (i.e., a value of one corresponds to a perfect matching between the measured and the DT computed values); and subsequently a max value of one for the cross validation set. A typical run of the model took about ten minutes on a 850 MHz PC machine, 128K RAM.

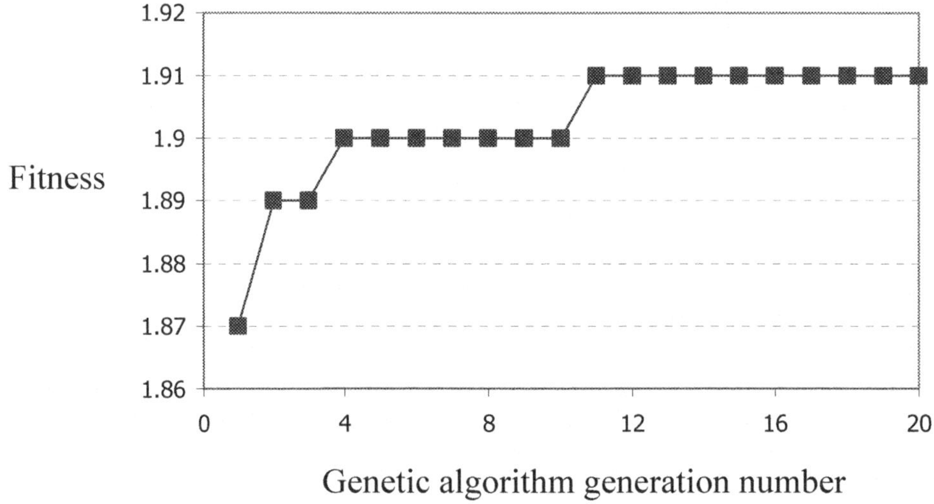

Fig. 5: Typical convergence of the Hybrid (DT-GA) Model

Figs. 6 and 7 show results for the quantity and quality models. Fig. 6 shows results for the cross validation data set for the period of 1996-97 for flow at the Meshushim outlet; Fig. 7 shows results for the cross validation data set for 1998-99, for total Nitrogen. In both cases the fitness was above 1.92 (see Fig. 5).

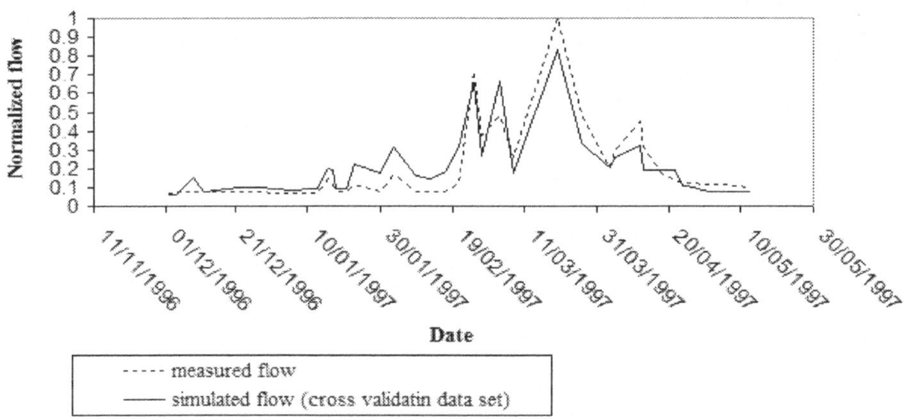

Fig. 6: Cross validation data for 1996-97: quantity section

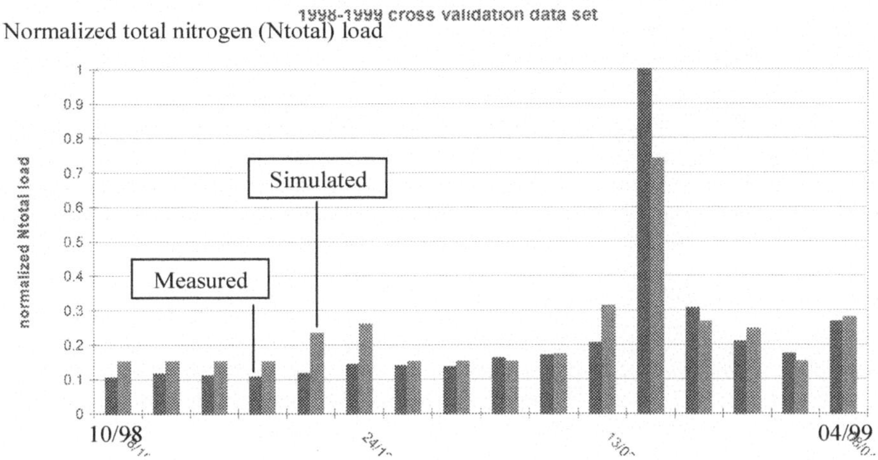

Fig. 7: Cross validation data for 1998-99 – total nitrogen (Ntotal): quality section

CONCLUSIONS

Data driven modeling (DDM) is the study of computer algorithms that improve automatically through experience. During the last decade DDM techniques became one of the more robust and widely used schemes in both research and practice. DDM applications range from data mining programs that discover general rules in large data sets, to information filtering systems that automatically learn users' interests.

The current model couples genetic algorithms with decision trees in an overall methodology for predicting runoff flow and contaminants in watersheds. The model was applied to a small watershed in northern Israel showing promising results for predicting both flow and contaminations as a result of rainfall storms.

ACKNOWLEDGMENTS

This paper is part of an ongoing project funded by the Israeli Water Commission, and by the Technion Grand Water Research Institute (GWRI).

REFERENCES

Abebe A. J., Solomatine D. P., and Venneker R. G. W. (2000). Application of adaptive fuzzy rule-based models for reconstruction of missing precipitation events. *Hydrological Science Journal*, Vol. 45, No. 3, pp. 425 – 436.

Castano M., Mejuto M. F., and Calera A. (1998). Application of remote sensing techniques to the evaluation of water consume. *The International Society for Optical Engineering*, Vol. 3499, pp. 58-68.

DHV MED (2001). *http://www.dhvmed.com/GISweb2.htm*

Dorado J., Puertas J., Santos A., Rivero D., Pazos A., Rabunal J. R. (2003). Prediction and modeling of the rainfall-runoff transformation of a typical urban basin using ANN and GP. *Applied Artificial Intelligence*, Vol. 17, No. 4, pp. 329-343.

Goldberg D. E. (1989). Genetic Algorithms in Search, Optimization, and Machine Learning. *New York: Addison-Wesley.*

Hewett R. (2003). Data Mining for Generating Predictive Models of Local Hydrology. *Applied Intelligence*, Vol. 19, No. 3, pp. 157-170.

Holland J. H. (1975). Adaptation in Natural and Artificial Systems. *Ann Arbor: The University of Michigan Press.*

Lohani V. K. and Loganathan G. V. (1997). Decision tree approach for drought management. *Proceedings of the Annual Water Resources Planning and Management Conference, Aesthetics in the Constructed Environment*, pp. 100-105.

Markel D. and Shamir U. (2002). Monitoring Lake Kinneret and Its Watershed: Forming the Basis for Management of a Water Supply Lake. In *Water Resources Quality Preserving the Quality of our Water Resources, Rubin H., Nachtnebel P., Fuerst J., and Shamir U. (Eds.), Springler – Verlag,* pp. 177-190.

Quinlan J. R. (1986). Induction of decision trees. *Machine Learning 1*, 1, pp. 81 – 106.

Quinlan J. R. (1993). C4.5: Programs for Machine Learning. *San Mateo, CA: Morgan Kaufmann.*

Solomatine D. P. and Dulal K. N. (2003). Model trees as an alternative to neural networks in rainfall-runoff modeling. *Hydrological Science Journal*, Vol. 48, No. 3, pp. 399 – 411.

Trauth R. and Xanthopoulos C. (1987). Non-point pollution of groundwater in urban areas. *Water Research*, Vol. 31, No. 11, pp. 2711-2718

AGNPS and SWAT Model Calibration for Hydrologic Modelling of an Ecuadorian River Basin under Data Scarcity

D. Matamoros*, E. Guzman*, J. Bonini* and P.A. Vanrolleghem**

* Facultad de Ingenieria Maritima, Escuela Superior Politecnica del Litoral, P.O Box 09-01-5863, Guayaquil, Ecuador. (E-mail: dmata@goliat.espol.edu.ec)
** Department of Applied Mathematics, Biometrics and Process Control. Ghent University, Coupure Links 653, B-9000 Gent, Belgium. (E-mail: Peter.Vanrolleghem@ugent.be)

Abstract Chemical fate assessment in a river basin requires flow and sediment characterization that can be performed by continuous field measurements or modelling. Once this step is completed, the environmental modelling can be started. In current assessments done in Ecuador, field data are not easily available and are also scarce. Therefore, the use of models is a challenge. Two models, AGNPS and SWAT, were tested to evaluate their potential applicability under data scarcity. After gathering all necessary data, the models were calibrated from a hydrological point of view. Some inverse modelling techniques were applied to overcome the lack of necessary data. This paper shows the results of the hydrological calibration of both models.

Keywords AGNPS, Model Calibration, SWAT, Watershed Assessment.

Problem Definition

The Flemish Interuniversitary Council (Belgium) and the Guayaquil Polytechnics School (Ecuador) are currently evaluating the potential environmental impacts that could occur in a river basin with intensive banana production. Two models able to simulate the impact of pesticide usage in a river basin were compared: AGNPS and SWAT. The selected watershed (Chaguana river) is located at the most south western part of Ecuador (Matamoros et. al, 2002). Before model usage, all necessary data were gathered, processed and converted in the appropriate format. Among the data collected, precipitation and flow information were the main concern in view of the hydrological calibration of both models.

The flow data was collected from three gauging stations (Chaguana and Zapote). The flow measurements represent average monthly flows and cover a period of 4 years (1979, 1980, 1982 and 1983). There is no available measurement after 1983. The location of gauging stations is depicted in Figure 1.1.

Although there was information from 8 surrounding weather stations, reliable precipitation data were only obtained from four of them (Machala, Pagua and Pasaje stations). Figure 1.1 also shows the location of weather stations. Precipitation data represent total monthly values and also the maximum 24-h precipitation in every month. The recording period is from 1979 to 1999.

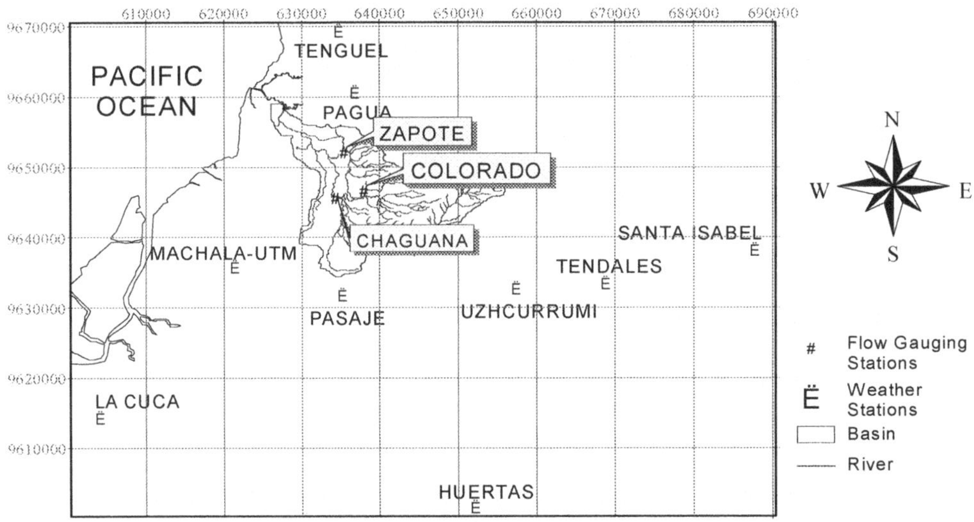

Figure 1.1 Location of weather stations and flow gauging stations for Chaguana basin.

The official database, where the flow data were obtained, did not contain measurements of suspended sediments to perform the model calibration. Therefore, thirteen monitoring points, including the place of gauging stations, were set along the river to determine suspended sediments and flow rates during four sampling campaigns between 2001 and 2002 (14 November 2001, 30 March 2002, 5 July 2002 and 11 November 2002). Due to accessibility issues, only 46 km out of 320 km of the river system were sampled. Each sampling campaign represents a single event for modelling purposes. Figure 1.2 gives the location of the sampling points along the Chaguana river basin.

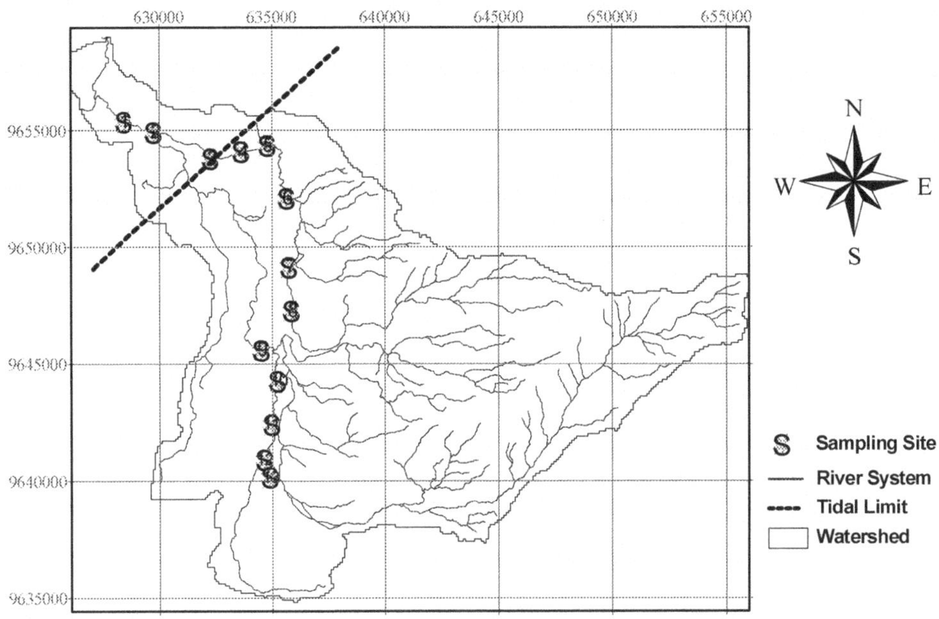

Figure 1.2 Location of sampling points along the Chaguana river basin.

Methodology

Basin Division

The research team that evaluated the AGNPS model divided the Chaguana river basin in 192 AGNPS cells which drain into 78 river reaches (see figure 1.3.a). An explanation of the meaning of an AGNPS cell can be found in the AGNPS User's manual and in Matamoros et al (2004). The resulting cell areas vary between 1 and 829 Ha.

The second research team (Bonini and Guzman, 2003) evaluated the SWAT model by dividing the basin in 44 sub-basins, mainly based on the locations of the three existing gauging stations (Colorado, Zapote and Chaguana gauging stations), as shown in figure 1.3.b. The SWAT division is based on the concept of Hydrological Response Units (HRUs) applied to each sub-basin.

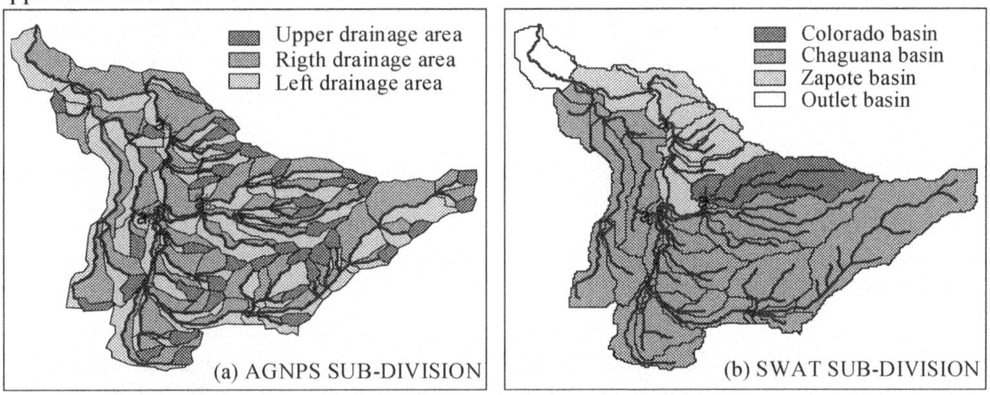

Figure 1.3 Chaguana basin division based on the applied model.

Fit Criteria

The whole calibration procedure was done with data collected from the three available gauging stations, and sediment samples gathered during the four sampling campaigns. Three statistical parameters were used to determine the goodness of fit of the predicted values related to the measured values:

- The Coefficient of Determination (r^2) is the square of the Pearson's Product Moment Correlation Coefficient, and it varies from 0.0 (poor model) to 1.0 (good model).

$$r^2 = \frac{\sum_{i=1}^{N}(O_i - O_{avg})(P_i - P_{avg})}{\left(\sqrt{\sum_{i=1}^{N}(O_i - O_{avg})^2}\right)\left(\sqrt{\sum_{i=1}^{N}(P_i - P_{avg})^2}\right)}$$

Where

O i , P i — Observed and Predicted value for each modelled event

O avg , P avg — Observed and Predicted average value for the evaluated range of data

- The Coefficient of Efficiency (E), developed by Nash and Sutcliffe (1970), ranges from minus infinity (poor model) to 1.0 (good model).

$$E = \frac{\left(\sum_{i=1}^{N}(O_i - O_{avg})^2\right) - \left(\sum_{i=1}^{N}(O_i - P_i)^2\right)}{\left(\sum_{i=1}^{N}(O_i - O_{avg})^2\right)}$$

- The Index of Agreement (*d*) developed by Willmott (1981) presents the same range of values as the coefficient of determination.

$$d = \frac{\left(\sum_{i=1}^{N}\left(\left|P_i - O_{avg}\right| + \left|O_i - O_{avg}\right|\right)^2\right) - \left(\sum_{i=1}^{N}(O_i - P_i)^2\right)}{\left(\sum_{i=1}^{N}\left(\left|P_i - O_{avg}\right| + \left|O_i - O_{avg}\right|\right)^2\right)}$$

For further information, Legates and McCabe (1999) have written a complete discussion on these three statistical coefficients normally used in hydrological and climatic model evaluations. In addition, a relative bias was estimated for every pair of measured and predicted values. Then, an average was estimated for all the sampled values based on the following equation.

$$Bias_{AVERAGE} = \frac{1}{N}\sum_{i=1}^{N}\left(\frac{O_i - P_i}{O_i} \times 100\right)$$

Flow Calibration

The first calibration step adjusted the flow predictions as close as possible to the flow measurements. As written before, three gauging stations, known as Chaguana, Zapote and Colorado gauging stations, were used. They are mainly located in the middle course of the river basin.

The models evaluated by the research teams are mainly runoff-based. One of the most sensitive parameters for flow calibration in runoff-based models is the SCS Runoff Curve Number, which is an indicator of how much water is running off from the soil surface. The higher the Curve Number, the higher the estimated runoff that is obtained. Therefore, flow calibration was conducted by adjusting the Curve Number for each land use type involved in the basin assessment.

In the models, flow estimates are based on single or continuous daily events (rainfall), so it is necessary to have daily data to calibrate the model. In the present research, the gauging stations only had average monthly flows. In addition, weather data were also limited to total monthly values, the number of rain events and the maximum 24-hour precipitation fallen during every month. Therefore, it was necessary to generate daily precipitation data for the recorded period of the gauging stations by considering the following assumptions:

1. There is only one maximum precipitation event in every month corresponding to the recorded 24-hour precipitation at a specific reported day.
2. As a first approach, the rest of the monthly precipitation is equally distributed among the recorded number of rainy days in a month.
3. The model is run for every estimated daily event for that month. A mean monthly flow is obtained by averaging the resulting daily flows. That average monthly flow is compared with the reported flow in the corresponding gauging station.
4. If the statistical parameters are still showing "poor" fit, then the daily events in the month are rearranged by always keeping the maximum monthly precipitation and the number of rainy days in mind. This process is repeated until the flow predictions fit the measured values.

Suspended Sediment Calibration

Both models predict suspended sediments based on precipitation events (single or continuous). However, there was no information regarding the precipitation that fell during the sampling campaigns. Therefore, it was necessary to use the flow prediction to estimate the unknown inputs (rain events) which are producing the sediment outputs for calibration purposes. This method is known as Inverse Modelling, and it has been used in several applications. Basically, Inverse Modelling is the use of a model output to estimate a model input.

The models were only run in the river reaches that showed measurements of suspended sediments during one of the sampling campaigns. The sediment calibration was performed by adjusting the two parameters that mainly contribute to the sediment yield and do have more uncertainty in their estimated values: the cover management factor (C) and the practice management factor (P).

Results and Discussion

Flow Calibration

Basically, the Colorado gauging station was calibrated first because it has a smaller drainage area and only two land cover types. The Zapote gauging station was calibrated second because it is located immediately downstream the Colorado station. And finally, the Chaguana station was calibrated by adjusting the curve number for each of its drainage basins. Unfortunately, there was no gauging station at the outlet of the basin. Table 1.1 shows the estimated statistical coefficients of fit in all three gauging stations for both model runs.

Table 1.1 Coefficients of fit during flow calibration for both models in the three gauging stations.

Gauging Station	Model	r^2	E	d	Bias (%)
Colorado	AGNPS	0.83	0.37	0.88	90.31
	SWAT	0.85	0.72	0.90	10.00
Zapote	AGNPS	0.85	0.53	0.91	26.79
	SWAT	0.88	0.77	0.94	-23.76
Chaguana	AGNPS	0.87	0.73	0.93	0.83
	SWAT	0.90	0.80	0.95	-8.79

The AGNPS model could predict the flows in the Chaguana gauging station fairly well (r^2 = 0.87, E = 0.73, and d = 0.93). For the Colorado and Zapote stations, the model showed lower values of goodness of fit (E = 0.53 for Zapote and 0.37 for Colorado). It is concluded that the AGNPS model usually fails to predict flows that occur in very small drainage areas with very low precipitation events, and this is mainly because the output results are restricted to three decimal place positions. Therefore, any predicted flow below 0.001 m^3/s (1 litre per second) is reported as zero. In addition, the lack of data for flow validation is critical; the data mainly represent extreme events (an "El Niño" event occurred during 1982 and 1983).

The outcome of the SWAT model was more accurate than the AGNPS model. The efficiency (E) of the flow calibration process was above 0.7 for all gauging stations. Although predictions for the Zapote and Colorado stations were improved, they are still below the ones for the Chaguana station.

The main reason for the result improvement is that the SWAT model uses the concept of Hydrologic Response Units (HRU) to couple land cover and soil information within each sub-basin. As described in the SWAT user's manual, Hydrologic Response Units are portions of a sub-basin that possess unique landuse, management and soil attributes. Although this concept is similar to the attribute of an AGNPS cell, the main difference is that a SWAT sub-basin can have many HRUs within it. On the other hand, AGNPS sub-basins can only contain a maximum of three AGNPS cells. Therefore, the information loss in SWAT can be reduced significantly when aggregating data in a sub-basin. This improvement results in better predictions because of a better characterisation of the evaluated basin. However, the improvement on the accuracy could be jeopardised if soil or land cover information is not accurate too.

Suspended Sediment Calibration

To perform the suspended sediment calibration, it was necessary to use flow predictions to assess the unknown inputs (rain events) which are needed to calculate the suspended sediment yield. This method is known as Inverse Modelling, and it has been used in several applications. Basically, Inverse Modelling is the use of a model output to estimate a model input.

In the present case study, the models required the precipitation on the campaign dates to predict the sediment yield. Due to the lack of precipitation data for the sampling days, it was necessary to use the inverse modelling technique. First, a flow graph was obtained from the flow calibration step (figure 1.4) for three locations in the river: the outlet, the Chaguana station and the Zapote station. The estimated rain events for the sampling days were obtained by introducing the monitored flow values in figure 1.4. The obtained rain event intensities represent an average rain event as falling at the same time in the entire catchment area. Table 1.2 gives the estimated rain event for the four sampling campaigns, which are the values used in the model simulation.

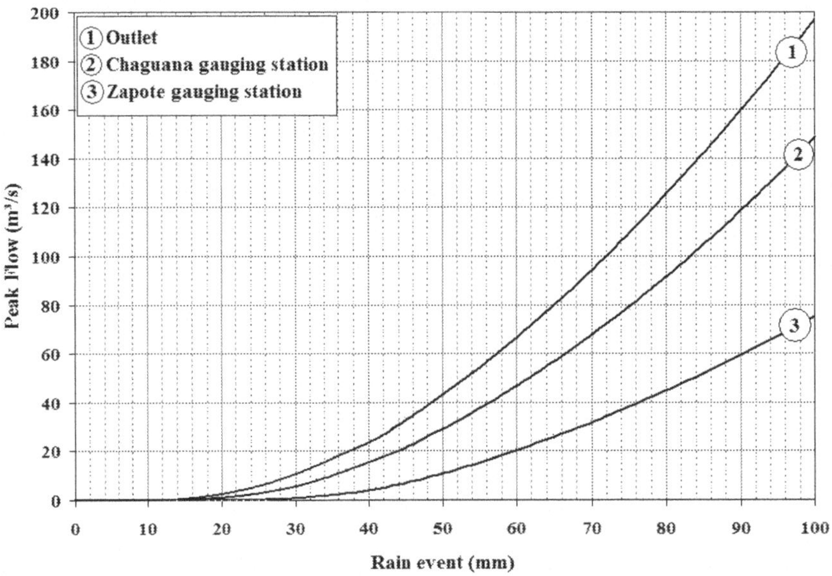

Figure 1.4 Estimated flows based on rain events for three locations in the river system.

Table 1.2 Estimated daily rain event for the sampling days based on the flow calibration graph.

Campaign	Date	Estimated Daily Rain Event
First	14 November 2001	4 mm
Second	30 March 2002	58 mm
Third	5 July 2002	3 mm
Fourth	11 November 2002	5 mm

The suspended sediment calibration procedure was explained in the methodology section. Basically, the cover management factor (C) and the practice management factor (P) were adjusted to fit the predicted values around the observed values. Table 1.3 gives the statistical coefficients of fit in the sampled reaches of the basin for both model runs.

Table 1.3 Coefficients of fit during suspended sediment calibration for both models in the main rivers of the Chaguana system.

Gauging Station	Model	r^2	E	d	Bias (%)
Zapote	AGNPS	0.60	-0.08	0.70	-30.37
	SWAT	0.96	0.86	0.96	-61.89
Chaguana	AGNPS	0.94	0.88	0.97	-3.64
	SWAT	0.97	0.92	0.98	-4.71

Regarding the AGNPS simulation, it is clear that the predictions for the Zapote River are not good enough as the coefficient of efficiency (E) is around zero. However the model can predict the sediment behaviour in the Chaguana river fairly well. There could be many reasons for this difference:

- The Zapote river has less reaches sampled than the Chaguana river during the sampling campaigns. Thus the characterization of this river is quite low.
- As said in the flow calibration, the model has a lower prediction efficiency in cells with small drainage areas. The Zapote river does not have as long a course (drainage area) as the Chaguana river, so the predictions are affected by this difference.
- Perhaps, the most important reason is the information loss that occurred during the spatial data aggregation done by the AGNPS model.

The SWAT model gave a better prediction in both sampled rivers (coefficient of efficiency above 0.8). This behaviour was expected as explained in the flow calibration process because the SWAT model minimizes the information loss by using multiple HRU's in each sub-basin.

Finally, for both models, there were also significant differences between predicted and measured values at specific sampling locations in the Chaguana river for the March sampling period (rainy season) while the rest of the sampling dates showed a good agreement. However, there was some dredging activity at certain locations along the Chaguana river between Reaches 45 and 22. This civil work affected the sediment yield by increasing the suspended solid concentrations on those reaches. For that reason, both models failed to predict the suspended sediment concentrations as the models calculate sediment yield based mainly on soil erosion from runoff. In addition, the last reach, Number 2, still shows predicted values lower than the actual ones. That is because that reach was also influenced by the tidal push at the moment of sampling. When the tide is entering the basin, some sediment is pushed back into the basin. This additional sediment load cannot be predicted by the runoff models. Therefore, both models are no longer applicable for those situations.

Conclusions

There were not enough data available for use in the modelling exercise. Thus, it was necessary to apply inverse modelling techniques to obtain certain unknown inputs. The use of inverse modelling represents a useful approach to overcome problems regarding input data. However, this technique also represents a risk because the outcome may not represent the phenomena to be modelled. For that reason, it is always better to use real data in the modelling process.

In the case study, the lack of daily rainfall data to run the model for the sampling events represented an opportunity to apply inverse modelling techniques. The obtained outputs were within the expected range.

The predicted flows were within acceptable ranges for the Chaguana gauging station, while predictions for the Zapote station are still not good. A potential cause can be that the Zapote river is not as well characterised as the Chaguana river.

Sediment predictions also show a reasonable agreement. However, there are two situations where the models cannot be applied: sediment supplied by civil works conducted in river banks and the sediment push-back caused by tidal influence.

Regarding model performance, the SWAT model with less basin subdivision showed to be more accurate than the AGNPS model. This is explained by the use of *Hydrological Response Units* (HRUs) in the SWAT model, which causes less information loss when doing data aggregation in the processing step.

Acknowledgement

This work has been supported and funded by ESPOL-VLIR project (Component 4). The authors would also like to thank the financial support of the EU Harmoni-CA project EVK1-CT-2002-20003.

References

Bonini, J. and Guzman, E. (2003) Flow and Sediment Estimation in Chaguana River Basin by using SWAT model. B.Sc. thesis, Escuela Superior Politecnica del Litoral, Ecuador (*in Spanish*).

Legates, D.R. and McCabe, G.J. (1999) Evaluating the use of "Goodness of Fit" measures in hydrologic and hydroclimatic model validation. *Water Resources Research*, **35**, 233-241.

Matamoros, D. (2004). Predicting river concentrations of pesticides from banana plantations under data-poor conditions. Ph.D. thesis, Ghent University, Belgium.

Matamoros, D., Bonini, J., Guzman, E., Ramirez, G., and Vanrolleghem, P.A. (2002). Design and implementation of a measuring campaign to model pesticide impacts in an Ecuadorian watershed. *Wat. Sci. Tech.*, **46**(9), 37-45.

Matamoros, D., Van Griesven, A., Van Biesen, L. and Vanrolleghem, P.A. (2004) Development of a Geographical Information System for Pesticide Assessment on an Ecuadorian Watershed. In *Proc. of the IWA World Water Congress and Exhibition*, Marrakech, Morocco 19–24 September (*in press*).

Simulation of flood flow over Bangkok flood plain by using artificial neural network

Tawatchai Tingsanchali*, Feixia Wang**

* School of Civil Engineering, Asian Institute of Technology, Pathumthani, Thailand 12120
**Same address as above

Abstract Artificial neural network (ANN) is an effective tool in simulating flood forecast in the eastern flood plain of Bangkok. Even though ANN does not require physical or hydrological understanding, its input data should be carefully analyzed and selected. Dense vegetation and land depression storages in the study area show a significant time-delayed effect on flood level hydrograph compared to fluctuating pattern of rainfall. To represent this time-delayed effect, a moving time-average rainfall is used for input to ANN in simulating flood forecast. Since the moving average rainfall requires rainfall forecast in real time flood forecast operation, a trade-off between time periods for rainfall forecast and for flood forecast should be properly balanced.

Keywords Artificial neural network, flood flow simulation, flood plain, flood forecast.

Introduction

Through many innovations of hydro-informatics and other such development, numerical-hydraulic models (Tingsanchali and Vongvisessomjai, S., 1986; Dibiki and Abbott, 1999) are now becoming even more accurate and reliable tools for the analysis, design, and management of a wide range of water resources projects. There are a number of applications, such as real time forecasting, and on-line control, for which such computing times are excessive. Therefore, the need arises to reduce the time and input data required by the hydraulic models to simulate hydraulic behavior of water resources systems.

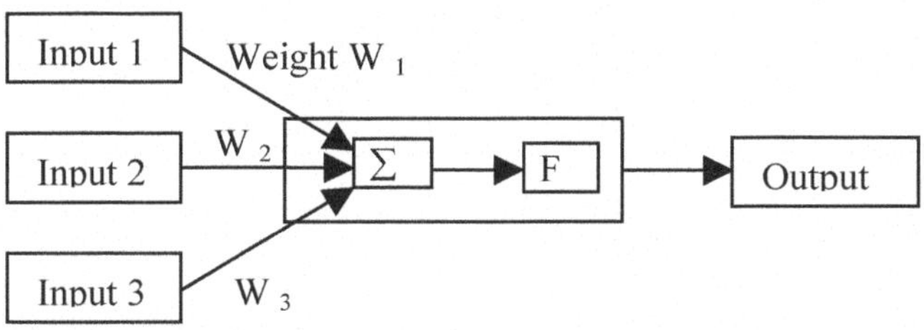

Figure 1 Artificial neural network with activation function F

This research deals with the application of artificial neural networks (ANN) (ASCE, 2000) with back propagation (Rumelhart, Hinton and Williams, 1986) to simulate flood flow in the eastern Bangkok flood plain area of 623 km^2 located about 20 km east of Bangkok.

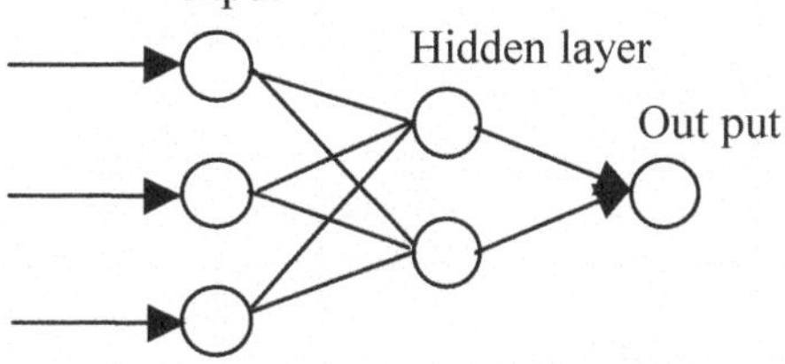

Figure 2 Multilayer feed forward network

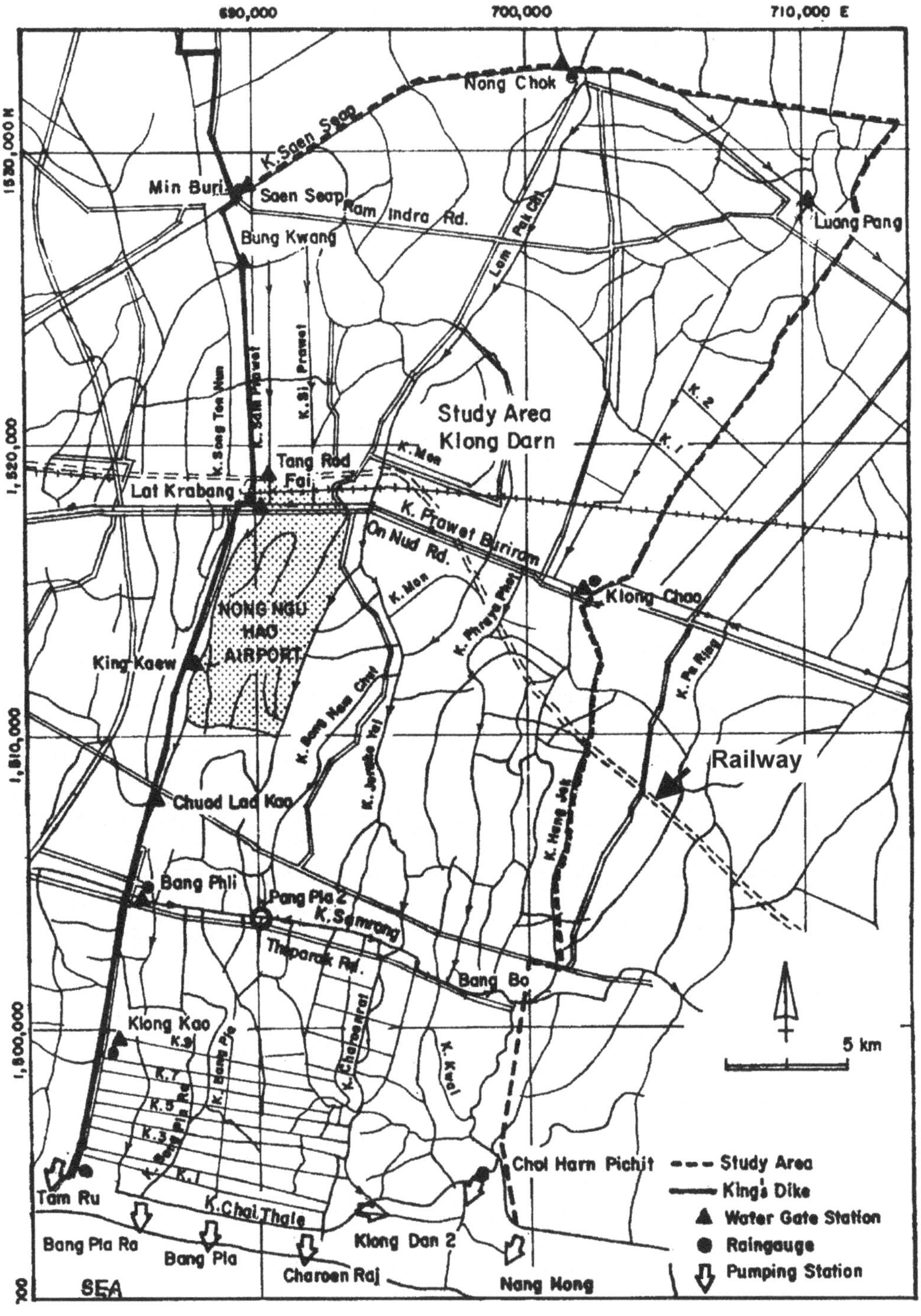

Figure 3 Study area, East Bangkok floodplain

Data collected in the large flood years in 1995, 1996 and 2001 are used to train and test the artificial neural network model to demonstrate ability of ANN for flow simulation in the flood plain. The performance of ANN is evaluated in term statistical performance indices on its accuracy and computational time. The effects of parameters of ANN on the computed results are also determined.

Description of Eastern Bangkok flood plain area

The flood plain area was developed from an original irrigation area (Klong Darn Irrigation project) to a residential region in the last 20 years. There exist a large number of irrigation canals, which overflowed during flood periods. The area is surrounded by flood control dikes with water control gates and is isolated from the Chao Phraya River in the west, the Bang Pakong River in the east, runoff from the north and the Gulf of Thailand in the south. There are five water level gages, seven rain gages, and six pumping stations along the main irrigation canals. Rainfall, water levels, and pumping discharges are recorded regularly on daily or hourly basis. In the east part of the study area, ground elevations are relatively low, however the lowest area is in the southwest where the area has a bowl-shape region, which makes it easily subject to flood damages. The main cause of flooding is due to heavy rainfall during flood seasons. To mitigate severe flooding and damages, big pumping stations were constructed in the south end near to the sea to pump floodwater out from the bowl-shape region. The Second Bangkok International Airport (Nong Ngu Hao Airport) is located in the study area.

Artificial Neural Network (ANN)

Several algorithms of ANN exist. However, back propagation which belongs to supervised learning algorithm that compares the actual output to the target or specified output and then readjust the weights backward in the network. The same input is specified to the network in the next iteration, so the actual output will be closer to the target output. One of the basic requirements of the BP training is that the transfer function be continuous and differentiable. The sigmoid logistic non-linear function (Minns and Hall, 1995) which fulfills the above requirement is used with the value ranging between 0 to1.The standard back propagation training algorithm (Rumelhart, Hinton and Williams, 1986) is as followed:

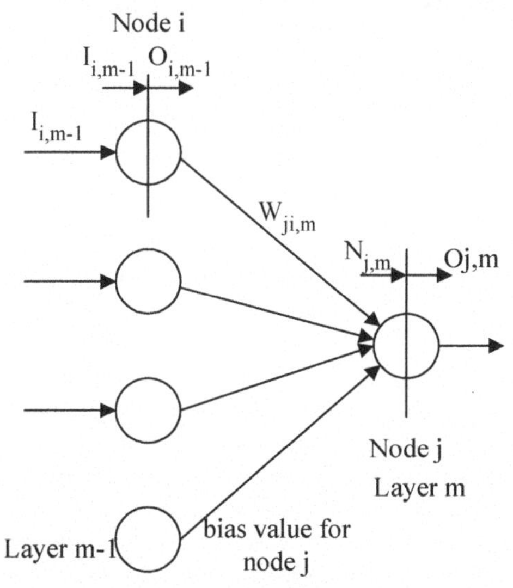

Figure 4 Transfer of input of layer m-1 to an output node of layer m

1. Initialize all weights and bias factors to small random values.
2. Forward pass: Present input vector $(I_1, I_{2,\dots,} I_{no})$ and specify the desired output $(t_1, t_{2,\dots,} t_{nl})$
3. For layer m = 1,2,…,l: , according to Figure 4, we can compute

$$N_{j,m} = \sum_{i=1}^{n_{m-1}} W_{ji,m} \cdot O_{i,m-1} + \theta_{j,m} \qquad (1)$$

where $O_{i,0}= I_i$, t_j = target value of neuron j in output layer, $O_{j,m}$= output of neuron j in layer m, $N_{j,m}$= activation of neuron j in layer m, $\theta_{j,m}$ = bias value for neuron j in layer m , $W_{ji,m}$ = synaptic weight between node j in layer m and node i in layer m-1

The output $O_{j,m}$ of the jth unit in the layer m is computed using a sigmoid function as

$$O_{j,m} = \frac{1}{1+e^{-N_{j,m}}}; j = 1,2,\ldots,n_m \tag{2}$$

4. Compute the final output $(O_{1,1}, O_{2,1}, \ldots\ldots, O_{n1,1})$ and compared with the desired output $(t_1, t_{2,\ldots}\ t_{n1})$. If the difference is acceptable, the process is terminated and the system has learned. Otherwise, continue to next step. When the number of epochs is reached while the difference is not acceptable, the convergence is not attained. One should try with a new set of initial values, or even modify the structure of the network.

5. Back pass: For layer m = l, l-1, l-2, …, 1, let $\delta_{j,m}$ = δ for neuron j in layer m,

For output layer m, $$\delta_{j,m} = O_{j,m}(1 - O_{j,m})(t_j - O_{j,m}); \tag{3}$$

For hidden layer m, $$\delta_{j,m} = O_{j,m}(1 - O_{j,m})\sum_{k=1}^{n_{m+1}} W_{kj,m+1}\delta_{k,m+1} \tag{4}$$

To compute the weight increments:

$$\Delta W_{ji,m}(n+1) = \eta \cdot \delta_{j,m} \cdot O_{i,m+1} + \alpha \cdot \Delta W_{ji,m}(n) \tag{5}$$

where η = learning parameter, α = momentum constant, $\Delta W_{ji,m}(n)$ = weight change between node j in layer m and node i at n iteration, $\Delta W_{ji,m}(n+1)$ = weight change between node j in layer m and node i at n+1 iteration, $O_{i,m-1}= I_{i,m}$, n = number of iteration (n = 1,2,3,…).The new values of the weight is computed as

$$W_{ji,m}(n+1) = W_{ji,m}(n) + \Delta W_{ji,m}(n+1) \tag{6}$$

where $W_{ji,m}(n)$ = weight value between node j in layer m and node i at n iteration, $W_{ji,m}(n+1)$ = weight value between node j in layer m and node i at n+1 iteration

6. Go to step 2.

Model training and testing

The training time is directly dependent on the size of the network, the larger the network size, the longer the training time. The training process depends on number of input units, number of hidden units and convergence criteria. To make the problem simple, a neural network with three layers is normally used, i.e., input layer, hidden layer and output layer. Too small number or too large numbers of hidden nodes pose difficulty to the training. Therefore the determination of the hidden layers and nodes is done by the trial and error. The basic way in which back propagation trained is that a set of input and corresponding output patterns is given in the training phase of back propagation. The system transfers the input to output based on initially randomly selected weights using the defined activation function.

The system output is then compared with the training set output and the difference between these two is what to be minimized in the so called “training phase” which is identical to “calibration” of the conceptual models. After training, the weights are kept unchanged and the neural network model is tested by applying it to other different sets of data. If the agreement is acceptable, the model is considered to be tested or verified. Otherwise more data is needed to train and test the model. The model is trained starting with some small randomly generated weights. After the predefined target error, e.g., 0.05 for 90% of all patterns is reached, the training process is stopped and the weights are saved. These weights and same NN architecture of the model are utilized in the testing phase.

In order to speed up the training process of ANN and avoiding oscillation of its output, the parameters called learning rate and momentum rate that control weight increments are adjusted. There is no specific rule in selecting the values of these parameters. However, adopting one set of values (for example, learning parameter α= 0.2 and momentum rate η= 0.5) the training process is started. Depending on whether the error reduction of the network output compared with the target output is slow or rapid and whether of the output of the network is oscillating or not, these two parameters are adjusted by trial and error.

Data collection

Different types of data such as daily water level, rainfall, and pumping discharge collected from the Klong Darn Irrigation Project (Figure 3) are applied to develop ANN model. The data collected are in the large flood years namely 1995, 1996, and 2001 during the flood periods from September to November. The daily data in 1995 and 1996 are used to train the model, and the daily data in 2001 is used for model testing. Flood flow in the study area has a two dimensional characteristic in horizontal plane.

Model input and architecture

The geometry of the study area is relatively flat except a bowl-shaped lower region. The study area is surrounded by dike system to prevent floodwater flowing into it. Due to diking, the fundamental model inputs would be rainfall and pumping discharges. Due to flat ground level and thick vegetation cover, the fluctuating effect of rainfall to water levels in the canals is damped out. The canal water level slowly responds to the rainfall patterns. The rainfall pattern is fluctuating irregularly with very short period while the water level during flood period normally will only has one peak in three months. The input rainfall pattern is not accordance with the output water level pattern. This will generate much error in model training and testing. Under this circumstance, one must understand the physical process of relationship of rainfall, water level and pumping discharge. Necessary data processing is required to achieve an output pattern corresponding to the rainfall. To represent proper output response, rainfall input is pre-processed to eliminate the fluctuating patterns to provide a moving time average value to ANN. Under this circumstance, the Moving Average Rainfall (MAR) is employed, for example, a three-day moving average rainfall MAR_i (Wang, 2002) is defined as $MAR_i = (R_{i-1} + R_i + R_{i+1}) / 3$, where R_i is the measured rainfall at the day i.

When the actual rainfall is used, significant fluctuation of computed flood water level is obtained. In reality rainfall after reaching the ground requires three to five days to move as runoff through vegetated flood plain and depression storages to reach the main canal. By trial

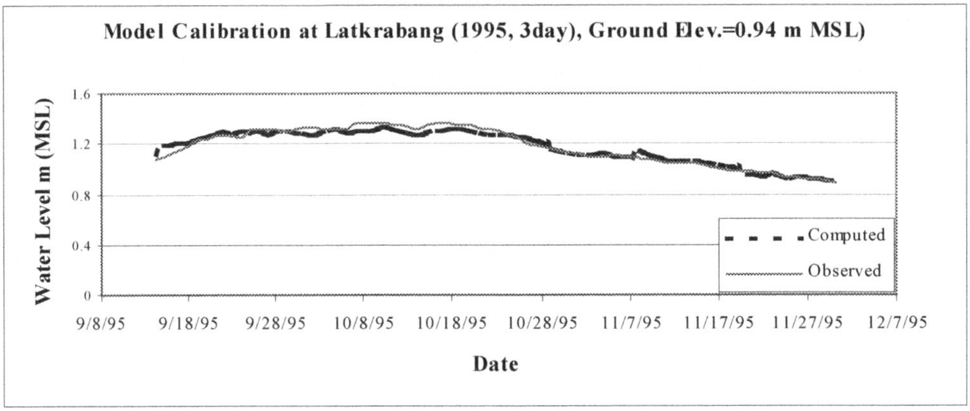

Figure 5 ANN model training for 3 day flood forecast at Latkrabang in 1995

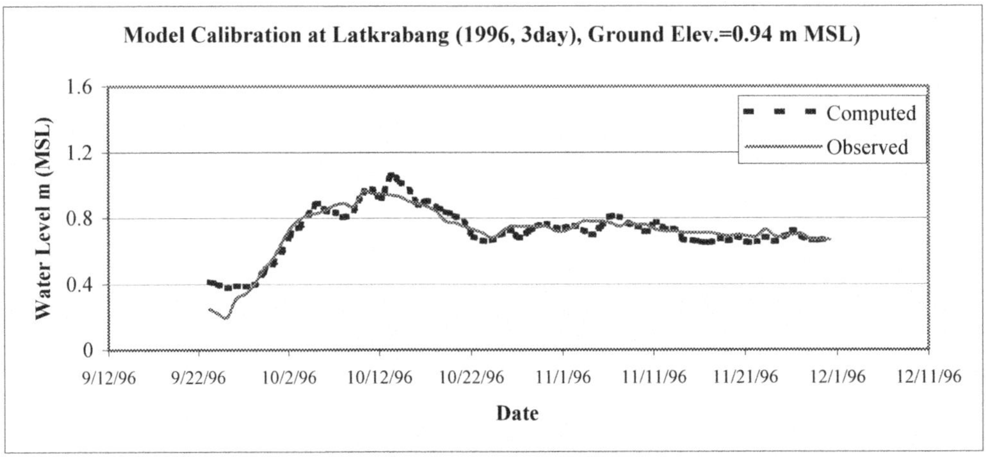

Figure 6 ANN model training for 3 day flood forecast at Latkrabang in 1996

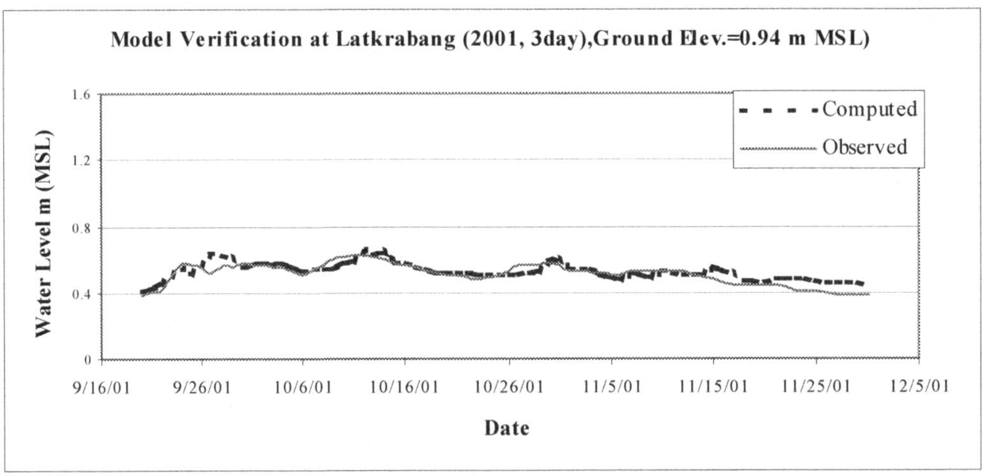

Figure 7 ANN model testing for 3 day flood forecast at Latkrabang in 2001

and error the suitable number of days for moving average rainfall is determined. The moving average rainfall gives significant improvement to the ANN training and testing.

By trial and error the ANN architecture is found. Neural network models of different architectures are developed based on the collected data. At Minburi station, the architecture 6-2-2-1 is represented by an input layer of 6 input variables, two hidden layers each of two hidden nodes and an output layer. At Latkrabang, the model architecture is 8-2-1. Usually the ANN architecture is kept unchanged for 1, 3 and 5-day ahead flood forecast.

Simulation results of ANN and discussions

When rainfall occurs, it will take time for the rainfall to flow into the canal, therefore, the effects of rainfall on the water level recorded along the irrigation channel would be damped and the flood levels rise and recede slowly. Because of the flat ground level and the vegetation over the floodplain, there will be three to five days delay of rainfall influence

By comparing the computed daily flood water level under the condition of actual daily rainfall, 3-day and 5-day moving average daily rainfalls, it is found that the 5-day moving average daily rainfall yields the most satisfactory result. For the 3-day moving average rainfall, in actual real time operation, the one day ahead rainfall forecast is required. Similarly, for 5-day moving average rainfall, two days ahead rainfall forecast is required. This study does not involve forecast of rainfall, from experiences it was found that the forecast of one-day or two-day ahead rainfalls can be done with acceptable accuracy. However forecast of more than three-day ahead rainfall is inaccurate. Therefore, no further attempt of using 7 day moving average daily rainfall is considered.

The results from ANN model training and testing are the comparison of the computed flood water levels with the observed data (Figures 5 to 7). Water levels at five water gages are simulated for flood forecasting for 1, 3 and 5 days ahead. Due to lack of field velocity measurement, no simulation of flood flow velocity is done. It is found that the flood level forecast at less advance time, i.e., 1 day and 3 day provide better results than those simulated from longer time step. The simulation results show that the accuracy of forecast is satisfactory for the three advance times. A trade-off between rainfall forecast for moving average and flood level forecast should be judged in real time operation for the most compromise solution.

Model performance evaluation

The model results are evaluated by statistical measures in terms of efficiency index EI and correlation coefficient R. The forecast accuracy decreases when the number of days in advance increases. From the results, it is clear that the NN model would yield satisfactory forecast of daily flood levels with time step 1, 3 and 5 days ahead in the study area. For example for 3 day ahead forecast at Minburi, the EI is 98.6 % for training and 87.6 % for testing while the corresponding correlation coefficient R is 0.993 and 0.96 respectively. For 1 day ahead forecast EI values are 99.7 % for training and 90.8 % for testing. For 5 day ahead forecast EI values are 93.5 and 78.1 % respectively.

Sensitivity analysis

Model sensitivity analysis is investigated for the effects of input data, learning parameter, momentum rate, model architecture and range of initial weights. Rainfall inputs have significant effects on model outputs. When the rainfall input is excluded, the model accuracy is significantly decreased, which shows its importance. In addition, the model forecasting ability is considerably improved when the moving average rainfall is employed instead of using the actual daily rainfall data.

It is found that assumed initial weights of ANN have certain effects on the optimal weights. Different sets of initial weights are used to assure the optimal results of ANN. In addition, more computational time is required when decreasing the target error. If the target error is reduced from 0.05 to 0.03, the computing time will increase from 1 minute to about 15 minutes. The smaller learning parameter will require more computational time because it is related to the change of the weights. It is found that the influence of the momentum rate is more significant in improving the convergence than that of the learning parameter and other parameters.

From the results, it is found that more hidden nodes do not always yield better model performance. In many cases simpler model architecture yields better than those complicated ones. Apparently, model produces better water level forecast for 1 day and 3 days ahead than 5 day ahead. In all cases the maximum absolute error in the flood level forecast simulation is less than 20 cm which is still acceptable in practice.

Conclusions

To ensure the model accuracy and reliability in this study, the input rainfall data to ANN model has to be pre-processed such as by the use of moving time average technique to represent the actual time delay in the flood water levels in the canals and in the field. Although ANN is not a physical-based model, the modeler must be aware of the physical situation of the simulation area such as ground slope, rainfall patterns, locations of water gages and pumping stations to under stand the characteristics of the study system. One should determine the appropriate and effective input for model training and testing.

It is practical and feasible to employ ANN model for the purpose of flood forecasting. NN model has demonstrated a strong capability to simulate flood flow over the eastern flood plain of Bangkok especially in the Klong Darn Irrigation project area. A trade-off between the number of days for rainfall forecast and daily flood level forecast should be properly balanced. A more comprehensive study should be carried out to get the most compromise solution.

References

ASCE, (2000). Task committee on application of artificial neural networks in hydrology., artificial neural networks in hydrology. ii: hydrologic applications. *Journal of Hydrologic Engineering*. **5(2),** 124-137.

Dibike, Y.B. and Abbott, M.B. (1999). Application of artificial neural network to the simulation of a two dimensional flow. *Journal of Hydraulic Research*. **37(4)**, 435-446.

Minns, A.W. and Hall, M.J. (1996). Artificial neural networks as rainfall-runoff models, *Hydrological Sciences Journal.* **41(3),** 339-417.

Rumelhart, D.E., Hinton, G.E. and Williams, R.J. (1986). Learning representations by back propagation errors. *Nature*, 323, 533-536.

Tingsanchali, T. and Vongvisessomjai, S. (1986). Bangkok flood protection, Chao Phraya 2, AIT*Research Report no. 186*, Conducted for Bangkok Metropolitan Administration, Thailand.

Wang, F. (2002). Simulation of two-dimensional flow over floodplain by using neural network, Master Thesis WM-01-6, Asian Institute of Technology, Pathumthani, Thailand.

The Relation between Nutrient Management in the Danube Basin and Eutrophication Problems in Receiving Black Sea Coastal Waters

M. Zessner, H. Kroiss and Ch. Lampert

Institute for Water Quality and Waste Management, Vienna University of Technology, Karlsplatz 13/226, 1040 Vienna, Austria, (E-mail: mzessner@iwag.tuwien.ac.at)

Abstract The project daNUbs "Nutrient Management in the Danube River Basin and its Impact on the Black Sea" deals with the relation of the management of nutrients in the Danube Basin, to emission and instream loads into and in the river system of the Danube Basin and the influence of nutrient discharges on the Western Black Sea coastal area. The Danube River basin is the second largest river basin in Europe and is characterized by an extreme heterogeneity of anthropogenic as well as natural factors. The paper explains the concept and shows interim results from the daNUbs project. In the frame of the workshop "River Basin Management Using Machine Learning" this contribution tries to explain the complexity of questions related to river basin management on an international scale and the variety of approaches needed to derive management strategies. Following main outcomes that can be used as basis for management decisions can be concluded up till now. The status of the Western Black Sea coastal area has significantly improved within the last 15 years. This is mainly caused by reduced nutrient inputs from the Danube River. Phosphorus, which is the limiting nutrient now, is of main importance in this respect. The reduction of nutrient discharges is mainly due to the economic break down in big parts of the Danube Basin. Future nutrient management in the Danube Basin has - at least - to aim for a stand still in discharges of nutrients to the Black Sea. This stand still of discharges has to be achieved in combination with a recovery of the economy and agriculture, which will tend to increase nutrient emissions.

Keywords daNUbs, diffuse pollution, nitrogen, phosphorus, river basin management.

Introduction

Worldwide it is recognized that river basin management is the key for sustainability in water and land use management. This task is already difficult for national river basins only influencing the national coastal areas. For trans-boundary river basins influencing multinational coastal areas it is even more important to base water quality management on sound scientific understanding of the relation between pollution and water quality. Successful strategies for sustainable development additionally need to be based on sound political and economic considerations.

Danube River catchment including Black Sea coastal area is an excellent example for such a situation: 13 countries with many different languages, religions and different economic strength join the Danube catchment. The ICPDR (International Commission for the Protection of the Danube River) and the ICPBS (International Commission for the Protection of the Black Sea) have been established in order to find management solutions which are then adopted by the member states for implementation.

The economic crisis, caused by the crash of the communist system in several eastern Danubian countries (EDC) resulted in dramatic changes in nutrient management in agriculture and industry which created the opportunity to study the cause effect relationship of a management change in full scale.

Basic Characteristics of Danube River Basin

The Danube River catchment is the second largest river basin in Europe. At its mouth to the Black Sea it has a mean discharge of roughly $6500 m^3/s$ which represents a very high dilution capacity for all anthropogenic pollution loads.

The whole Danube basin has an area of ~ 800,000 km². The total length of the main river is 2,857 km. ~ 85 Mio. inhabitants are living in the catchment, the population density therefore is little more than 100 inhabitants/ km². There are 18 states which contribute to the catchment, 13 of them with more than 2000 km² (Austria; Bosnia-Hercegovina; Bulgaria; Croatia; Czech Republic; Germany; Hungary; Moldova; Romania; Serbia; Slovakia; Slovenia and Ukraine).

Table 1 shows the strong differences in economic strength in the Danubian countries. The gross domestic product per inhabitant in this table is already related to the purchase power parity. These data cannot easily be compared with developing countries in other regions, as they are strongly influenced by the crash of the communist economy 15 years ago.

Table 1 Economic strength of Danubian countries (IFIP, 2003)

Country Year 2002	GDP/capita EURO, **PPP***	In % of EU – 15
EU – 15	24,5	**100**
Austria	26,7	111
Germany	24,5	102
Czech Republic	14,6	61
Hungary	12,6	52
Slovak Republic	12,3	51
Slovenia	17,0	71
Bulgaria	7,7	32
Croatia	9,2	38
Romania	6,1	25
Bosnia&Herzegovina	6,5	27
Serbia&Montenegro	4,1	17
Ukraine	4,5	19
Moldova	2,1	9

* PPP = Purchasing Power Parity

Nutrients: a challenge for trans-boundary management

One of the major trans-boundary problems of a catchment is the land – ocean interaction, which can only be managed on a catchment scale. Within this context, nutrients (nitrogen and phosphorus) play a dominant role related to trans-boundary pollution:

- Nutrients strongly affect the marine coastal areas by eutrophication.
- This can be the case even river quality standards are met.
- Pollution from diffused and point sources are important.

- Storage capacity of nutrients in the catchment is orders of magnitude higher than yearly turnover.
- The whole population is contributing to the problems by their life style.
- Economic development or crises have strong impacts.
- The complex relations between anthropogenic and natural processes effecting nutrient fluxes makes understanding of cause-effect relationships difficult.
- The optimal application of scientific knowledge for management can only be achieved by the development of complex modeling tools.
- The development of management tools has to take into consideration the specific regional, political, economical and natural conditions.

Most of the topics mentioned above are addressed in the project daNUbs "Nutrient Management in the Danube Basin and its Impact on the Black Sea" which started in 2001 and will be finished in 2005. The project is coordinated by the Institute for Water Quality and Waste Management, Vienna University of Technology. The research team consists of 17 partner institutions and represents the work of more than 35 researchers. ICPDR is an active partner in this project. The daNUbs-project investigates the complex relationship between the anthropogenic point and diffused nutrient fluxes, the loads transported by river Danube and the reaction of the marine costal area. It aims at developing a sound basis for political and economic decision making. Due to the high dilution capacity of River Danube the application of an environmental quality approach for the river in all riparian countries alone would not guarantee a sustainable quality of the coastal waters of Romania and Bulgaria. All the following results are related to this project. This paper aims to show the complexity of aspects that have to be considered in context of river basin management on an international scale.

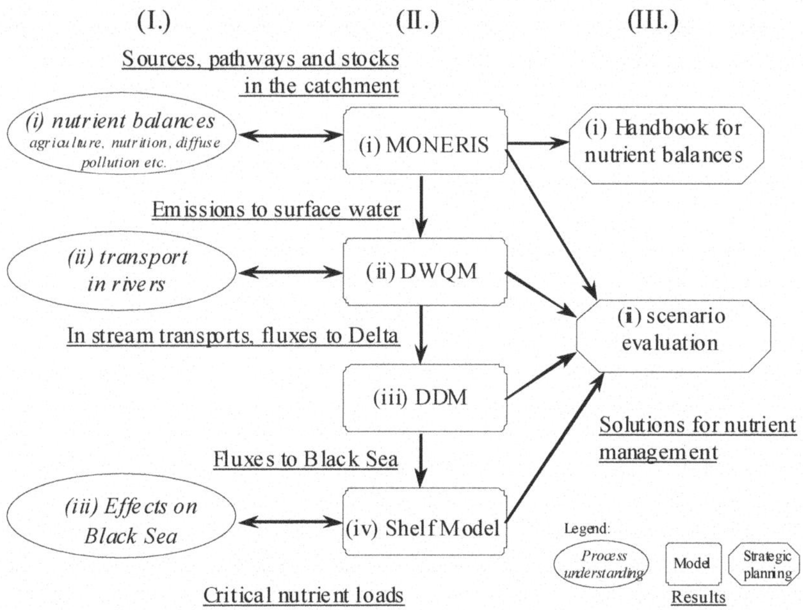

Figure 1 Basic concept of daNUbs project

Figure 1 shows a scheme of the basic concept of daNUbs project. Research elements of three different focuses are included: improvement of process understanding, improvement and application of mathematical models and considerations of strategic planning of nutrient management. A combination of research on improved process understanding and quantitative modeling in respect to sources, pathways and stocks of nutrients in the catchment, emissions to surface waters, instream transport and retention in the river system and effects on the Black Sea is used as basis for scenario evaluation for possible solutions of future nutrient management.

Reliable understanding of cause relationships between natural and anthropogenic processes and the development of mathematical models for quantitative description are not possible if they cannot be based on reliable, complete and correct data. This turned out to be a crucial point of investigations on such large scale and complex systems.

The correct interpretation of measured concentration data can only be achieved with understanding of the processes involved and the ability to describe their interactions adequately by models. If the data are not reliable the model calibration will be incorrect or at least erroneous which in turn can result in misleading interpretation.

Analysis of the water quality situation in Danube catchment

Danube is the main contributor to eutrophication phenomena in Western Black Sea coastal area. It discharges about 65 – 80 % of the nutrient load input to this part of the Black Sea.

The marine ecology in the Western Black Sea coastal area (WBSC) has significantly improved since the late eighties and early nineties. The improvements can be related to the following indications:

- Reduced eutrophication (algae production),
- Zoo-benthos and phytoplankton have recovered,
- N/P ratio (Redfield ratio) in WBSC is nearly optimal.
- The anaerobic conditions at the bottom of WBSC, indicators of eutrophication, have nearly completely disappeared in 1999-2002. They have shown a steady increase from about 1970 up to about 1990. The actual situation is similar to the 1960s.
- Only the fish stock is still out of balance. This is probably a result of over fishing. (Horstmann 2004)

The main causes of the improvements in WBSC are:

- Phosphorus loads transported by Danube have decreased by more than 30 % as compared to the situation in the eighties (Behrendt *et al.*, 2004). (see figure 4)
- Phosphorus is the limiting nutrient for algae growth in WBSC (Horstmann, 2004)
- Nitrogen loads from the Danube Basins to the WBSC have decreased as well (by ~20 % as compared to 1990, Behrendt *et al.*,2004). Despite dramatic changes in N-fertilizer application this changes are not very significant, because big nitrogen stocks (soil, groundwater) in the catchment slow down effects on diffuse N-emissions to the river system.

Reduction of nutrient discharges to the Black Sea by River Danube were caused by:

- Economic crises in the many Danubian Countries, which resulted there in
    - Closure of large industrial animal farms with manure discharges to the river system,
    - Reduced fertilizer production related with reduced emissions from fertilizer factories,
    - Reduced animal stock and density,
    - Strongly reduced market fertilizer application,
- introduction of P free detergents in Germany and Austria and the start of a similar development in the eastern Danubian countries,
- P and N removal at treatment plants (A, D,CZ),
- Progress in improved agricultural practice in Austria and Germany.

Possibilities for future nutrient management

Management of nutrients is only possible if the sources of the nutrients as well as their fate in the catchment are known. For this purpose the flows, transformations and storage processes in terms of mass loads have to be identified and described. The transformation and storage processes from the sources to the large rivers have been described using the MONERIS model, developed by Behrendt *et al.* (1999) and adapted and applied for Danube catchment by Schreiber *et al.* (2003). The transformation of the nutrient loads along the large rivers is modelled using the Danube Water Quality Model and the Danube Delta Model (van Gils and Constantinescu, 2002). The output data of Danube Delta Model are used by two ocean models, a physical model describing the flow of Danube water in the WBSC and an adapted quality model (Kourafalou *et al.*, 2004).

The main results of the MONERIS model calculations are shown in figures 2 and 3. Despite severe problems with data reliability for nitrogen compounds in River Danube the following statements can be made (Schreiber *et al.*, 2003): The actual N emission to the river system of Danube Basin is ~ 690 kt/year. The natural background (N losses without any anthropogenic influence) represents about 25% of the total N-emission to river system of the Danube Basin. It has the same magnitude as the N-discharge caused by waste water management (mainly point sources). Another contribution of about 25% can probably be influenced by improved agricultural practice (e.g. reduced intensity). The rest of the agricultural contribution can hardly be influenced if agricultural production is accepted. About 8% can be related to air pollution by combustion and traffic mainly with NO_x. About 65% (450 ktN/a) of the nitrogen emissions to the river system are discharged by river Danube to the Black Sea. The rest of the emissions is removed from the river system by denitrification. The river network of smaller rivers is much more effective in this respect as the Danube (including its main tributaries) as well as the Danube Delta, which contribute to these losses with only about 2 % each of the total emissions (van Gils, 2004).

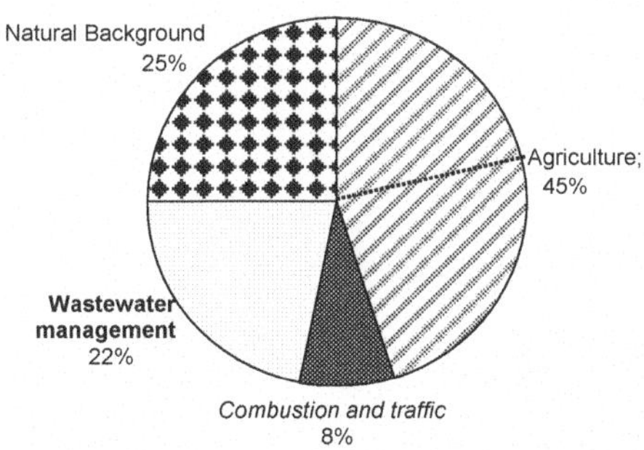

Figure 2 Sources of nitrogen emissions 1998-2000 into river system of the Danube Basin (Schreiber *et al.*, 2003)

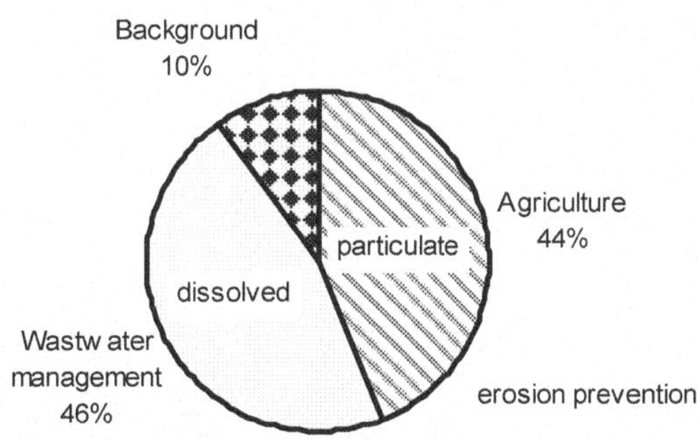

Figure 3 Sources of phosphorus emissions 1998-2000 into the river system of the Danube Basin (Schreiber *et al.*, 2003)

For Phosphorus the situation is different (figure 3). The yearly TP emissions to surface waters in the Danube Basin is actually about 70kt. Waste water discharges (point sources, combined sewer overflows and emissions from non sewered areas) contributes to about 45% of total and probably to roughly 65 to 75% of the dissolved easily accessible P-emissions to the river system of the Danube catchmetn. About 44% of total P-load stems from agricultural land and to a high extent is caused by erosion and is therefore in particulate form. The natural background is only about 10%, air pollution is negligible. About 35 % (about 25 ktP/a) of the emissions to surface waters reach the Black Sea (van Gils, 2004). The rest is stored in the rivers and the river corridor of the catchment. The region of the Iron Gate reservoir plays a significant role as point sink in this respect (storage of about 8-10 ktP/a) (van Gils, 2004). Nevertheless the main part of the storage of emitted phosphorus takes place in the river net of small to medium size rivers. Danube Delta contributes only with about 0.7 ktP/a to this retention (van Gils, 2004).

The regional and national contributions of agriculture to the trans-boundary transport to WBSC strongly differ as they depend on:

- Agricultural practice and volume of agricultural production (intensity of crop production, density of animals)
- Regional geologic, hydrologic and climatic conditions

Despite the fact that the actual situation shows important indicators of improvement in WBSC there are important risks for not reaching good ecological status in respect to WBSC eutrophication in the foreseeable future. This risk is linked to a recovery of the economy in the Eastern Danubian Countries (EDC) at the expense of environmental pollution. Special attention will have to be paid on the economic and especially to the agricultural development as the actual situation is economically not sustainable. In addition sewer development and recovery of industrial production (especially fertilizers) might lead to an increase of nutrient (phosphours!) discharges with negative effects on the Black Sea ecosystem. As figure 4 shows alone the implementation of EU urban waste water directive (UWWD) in respect to connections to sewer systems for all settlements > 2000 inhabitants might lead to a significant increase of P-emissions and discharges to the Black Sea coastal area, if treatment with P-removal (requirements for sensitive areas) is not implemented.

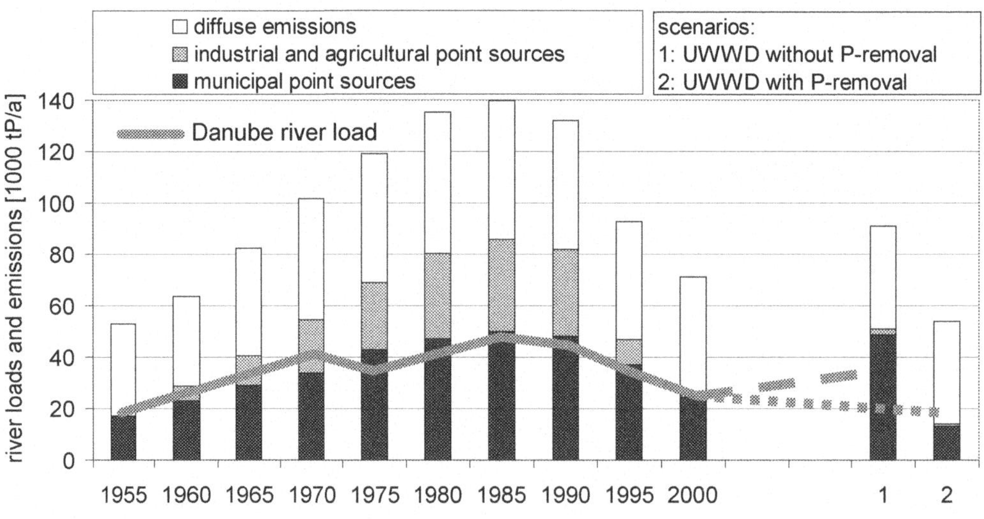

Figure 4 Development of phosphorus emissions and river loads (Danube discharges to the Black Sea) in the Danube catchment 1955 - 2000 (adopted from Behrendt *et al.*, 2004) and scenarios for future development.

Conclusions

Analysis of nutrient balances on catchment scale as basis for improved management is a highly complex subject. The daNUbs-project tries to combine a set of modeling approaches in order to cover the different aspects of nutrient emission, transformation and their effects on the marine ecosystem.

From the analysis of the actual situation and the historic development the following conclusions can be drawn for future nutrient management:

- A "stand-still" scenario has to be aimed at for the nutrient loads to the Black Sea but not for the economic development especially in the eastern Danubian countries.
- Economic development in these countries is desired even it leads to a certain increase of nutrient emissions (e.g. from agriculture in some regions)
- This increase of nutrient emissions has to be compensated by reductions at other sources.
- A development in agriculture similar to the period 1970 to 1990 has to be avoided, e.g.:
    - Redevelopment of centralized (industrialized) animal production,
    - Strong intensification of agricultural production by excessive market fertilizer application or fodder import,
    - Strong export oriented production (of animals).
- Strict control of phosphorus and nitrogen emissions from point sources in the whole catchment should be applied. This can be achieved by:
    - Implementation of EU Urban Waste Water Directive (271/91) using the requirements for "sensitive areas", as it is the case in Austria and Germany,
    - Introduction of P-free detergents,
    - Implementation of ICCP Directive for industrial and trade effluents with special emphasis on nutrient control.
- Development of sewerage (fulfilling EU UWWD) has to be strictly bound to adequate waste water treatment with nutrient (N and P) removal, or alternative sanitation systems with nutrient recovery will have to be applied. Actually only about 50% of the population in eastern Danubian countries is connected to sewer systems: From the nutrient management point of view sewerage should have a low priority; this is not the case from a ground water protection point of view.
- Improved nutrient management in agriculture (diffuse sources) has to be achieved by improved agricultural practice throughout the basin taking into account economical and ecological considerations but also the specific local climatic, geological and hydrologic situation.
- Erosion prevention is important to reduce the input of particulate phosphorus into the river system and the Black Sea: Particulate phosphorus is not immediately available for algae growth but represents a potential P-source for eutrophication.

Regarding transformation and storage of nutrients the following conclusions can be drawn:

- Measures for reduction of N emissions from diffused sources need long periods of time before they show effects because agricultural soils contain N loads which are about 100 times the yearly turnover of nitrogen. The same behavior is with particulate P from erosion.
- Retention (sedimentation) and losses (denitrification) of nutrients in the catchment depend on the specific local situation:
    - Small river systems with good (natural) river morphology (wetlands) have an important potential for nutrient retention and loss.
    - River Danube itself and its main tributaries play a minor role for nutrient retention.
    - Danube Delta and wetlands along the large rivers are not very relevant for nutrient retention (main flow through main channels)

  - Iron Gate Dam is a major sink for particulate phosphorus (actually about 15% of emissions)
- Nutrient management needs a long lasting strategy for sustainable development with a prospective of about 30 years for stable success.
- Data quality and monitoring improvement are needed in respect to correctness, reliability and completeness (e.g. TP, TN).

Acknowledgements

To the daNUbs Team: Results presented are a compilation of results elaborated by researchers of the daNUbs team. To the EU 5th framework programme: The results presented come to a large extent from the project "Nutrient Management in the Danube Basin and its Impact on the Black Sea" (daNUbs) supported under contract EVK1-CT-2000-00051 by the Energy, Environment and Sustainable Development (EESD) Programme of the 5th EU Framework Programme. Details on the project can be found on the project homepage: http://danubs.tuwien.ac.at/.

References

Behrendt, Huber, Kornmilch, Opitz, Schmoll, Scholz und Uebe (1999) Nährstoffbilanzen deutscher Flußeinzugsbiete, Forschungsbericht 296 25 515, Umweltbundesamt Berlin.

Behrendt, H., van Gils, J., Schreiber, H., Zessner, M. (2004) Changes of nutrient emissions from point and diffuse sources and loads in the Danube river system within the last decades, Archiv Hydrobiol., accepted.

Cosiasu A. (2004) daNUbs results cited in Horstmann, U. and Davidov, A. (2004) Effects of Reduced Danube Nutrient Discharge on the Northwestern Black Sea Ecosystem, ICPDR – UNDP/GEF workshop documents: Nutrients as a Transboundary Pressure in the Danube River Basin. 26 – 27 January 2004, Sofia, Bulgaria.

IFIP (2003) Historical economic development of the countries in the Danube River basin, part 1 of deliverable D9.1 of the EU-funded Project daNUbs, "Nutrient Management in the Danube Basin and its Impact on the Black Sea, contract EVK1-CT-2000-00051 by the Energy, Environment and Sustainable Development (EESD)

Horstmann, U. and Davidov, A. (2004) Effects of Reduced Danube Nutrient Discharge on the Northwestern Black Sea Ecosystem, ICPDR – UNDP/GEF workshop documents: Nutrients as a Transboundary Pressure in the Danube River Basin. 26 – 27 January 2004, Sofia, Bulgaria.

Kourafalou V, Staneva J. and Tsiaras K. (2004) Report on realistic simulations performed by model HB and 3-D coupled physical and biogeochemical model calibrated and validated for the Black Sea, deliverable D8.7 and D8.8 of the EU-funded Project daNUbs, "Nutrient Management in the Danube Basin and its Impact on the Black Sea, contract EVK1-CT-2000-00051 by the Energy, Environment and Sustainable Development (EESD)

Schreiber H., Constantinescu L., Cvitanic I., Drumea D., Jabucar D, Juran St., Pataki B., Snishko S., Zessner M. and Behrendt H. (2003) Harmonised Inventory of Point and Diffuse Emissions of Nitrogen and Phosphorus for a Transboundary River Basin, Research Report 200 22 232, on behalf of the federal environmental agency, Berlin, Germany

Van Gils J. and Constantinescu A. (2002) Draft Danube Delta Model and Danube Water Quality Model, first applicatons, deliverable D5.2 and D5.6 of the EU-funded Project daNUbs, "Nutrient Management in the Danube Basin and its Impact on the Black Sea, contract EVK1-CT-2000-00051 by the Energy, Environment and Sustainable Development (EESD)

Van Gils (2004) Transport and Retention of Nutrients in the Danube River and its Large Transboundary Tributaries, ICPDR – UNDP/GEF workshop documents: Nutrients as a Transboundary Pressure in the Danube River Basin. 26 – 27 January 2004, Sofia, Bulgaria

Author and Keyword Index*

*Authors and terms are indexed by the first page of the paper in which they appear.

IWA Publishing's authorised EU representative for General Product Safety Regulations is Diane D'Arras, 15 rue Duret, 75116 Paris, France, e-mail: safety@iwap.co.uk.

Printed and bound by CPI Group (UK) Ltd, Croydon, CR0 4YY
13/05/2026
02109623-0008